D0099729

*Environmental Damage
and Control in Canada*

2
A Citizen's Guide to Air Pollution

Canadian Society
of Zoologists

*Environmental Damage
and Control in Canada*
M. J. Dunbar, GENERAL EDITOR

1. **Environment and Good Sense**
M. J. Dunbar

2. **A Citizen's Guide to Air Pollution**
David V. Bates

David V. Bates

A CITIZEN'S
GUIDE TO
AIR
POLLUTION

SPONSORED BY THE CANADIAN SOCIETY OF ZOOLOGISTS
McGILL-QUEEN'S UNIVERSITY PRESS
MONTREAL AND LONDON
1972

© McGill-Queen's University Press 1972
ISBN 0 7735 0144 4 cloth; ISBN 0 7735 0145 2 paper
Library of Congress Catalog Card No. 72 75504
Legal Deposit Second Quarter 1972
Printed in Canada by John Deyell Limited

To my family

CONTENTS

PLATES

FIGURES

TABLES

PREFACE

Existing texts on air pollution may be divided into two broad classes, those that are primarily technical monographs or books containing detailed summaries of existing knowledge, and those which are in the general category of polemic texts. It is unfortunate that the cost of the former is often prohibitive and the technical accuracy of the latter is not infrequently questionable.

Proper legislative controls of air pollution must be built on the basis of an informed public opinion. Efforts to reduce significantly the present burden of air pollution are bound to have an effect on economic standards and on personal convenience, and it follows that decisions can only be taken in the presence of a general understanding of the reasons for adopting particular legislation at a particular point of time.

This short book has been written to act as a primary text for the many individuals who have a special responsibility in relation to air pollution. It is to be hoped that engineers, physicians, school teachers, and businessmen in general will acquire enough background knowledge on this contemporary problem to be able to understand the issues with which it deals. I do not share the opinion, expressed by some scientists concerned about environmental pollution, that it is necessary deliberately to exaggerate if any impact on the public or the politicians is to be achieved. If such a policy were to become general amongst the scientific community, far more would have been lost (the essence of science itself) than would have been gained.

The text has not been written with the conviction that air pollution is our most pressing public problem, nor in the belief that the health risks associated with it are in any way comparable to the enormous burden of disease occasioned by heavy cigarette smoking. Nevertheless, it has become increasingly clear that unless problems of air pollution are dealt with now they can only get worse in the future, and since the legislative and control processes will inevitably be slow there is every reason to urge that the present time is an appropriate one to begin to deal with air pollution. The text affords the reader every opportunity to delve into technical monographs and original sources of material, but such references are not by any means exhaustive. Finally, the reader will find that he is not given pre-packaged answers to the question of whether air-pollution legislation is required, but he may be able to understand the components of the question as a basis for formulating his own answer.

NOVEMBER 1971   DAVID V. BATES

M.D. (CANTAB.), F.R.C.P. (C.), F.R.C.P. (LOND.)

*Professor of Experimental Medicine and
Chairman, Department of Physiology, McGill University;
Senior Physician, Royal Victoria Hospital, Montreal.*

ACKNOWLEDGEMENTS

I wish to record my gratitude to the members of the McGill Interdisciplinary Committee on Air Pollution for the contributions they have made individually and collectively to my understanding of the different dimensions of this particular environmental problem.

I am also most grateful to M. Jean Marier, Director of the Air Pollution Division of the City Health Department of Montreal, for the many interesting and informative discussions we have had over the past six years on air pollution and its control.

I wish to thank the members of my department for their tolerance of the time I have devoted to studying different aspects of this problem, and, during the past year, for the fact that they have allowed my public-service teaching outside the University to be regarded as a substitute for a normal undergraduate teaching load in the department.

I am greatly indebted to Mr. Karl Holeczek of the Department of Physiology for making many of the figures and photographing them for reproduction. The preparation of the manuscript has been very capably handled by Miss M. Olejar and Mrs. Ruth Locke, both of the Department of Physiology.

I am grateful to the Graduate Faculty and the Principal of McGill University for financial assistance that made possible the inclusion of the four colour plates. The preparation of the volume has been greatly facilitated by the efficiency of the staff of McGill-Queen's University Press.

Finally, I wish to acknowledge the help I have received from the many students and members of the public who, by persistently asking difficult questions, have prompted me to consider the nature of some of the answers. Without their expressed concern there would have been no possibility of persuading the politicians to deal with air pollution until intolerable levels of some contaminants had been reached in some Canadian cities. Even so, as the Duke of Wellington remarked of the Battle of Waterloo, it is likely to be "the nearest run thing you ever saw in your life."

PLATE 1. Montreal at eight o'clock on an October morning
A temperature inversion commonly occurring between 2:00 A.M. and 7:00 A.M. results in a concentration of particulate and gaseous pollutants up to an altitude of about 500 feet. It is visible as a brownish-coloured pall over the city. Photograph by Professor T. Oke of the Department of Geography, McGill University.

PLATE 2. Montreal at eight o'clock on a February morning
The temperature was 8° below zero Fahrenheit when the photograph was taken. The sun, behind the tree in the foreground, is shining at the photographer. The lack of definition in the buildings is caused by the scattering of light by particles between the observer and the objects. At this temperature there is very little water vapour in the air. Photograph by the author.

PLATE 3. Particulate material in the lung of an urban dweller
A small airway, at left of centre, terminates in air sacs or alveoli. Particle deposition has occurred principally in the alveoli closest to the small airway, which would be about a millimetre or less in diameter. The patient from whom this specimen came was a tailor by trade, had no occupational exposure to dust, and was not a heavy smoker. It is not known whether this kind of particle deposition is injurious to the lung. Photograph by Professor W. M. Thurlbeck of the Department of Pathology, McGill University.

PLATE 4. Effluent plume from an industrial plant
In some industries, the problem of control of particulate pollution is formidable. This plant manufactures ferro-silicon by fusing rock and iron in a high-temperature electric arc. The effluent is silica, probably mostly in the amorphous form, with a particle size of less than one micron; the volume produced is believed to be in excess of 20 tons every 24 hours. The effluent plume can be photographed extending for fifteen miles downwind of the plant. The nature of the process makes the problem of trapping this volume of small particles, and disposing of them, a particularly difficult one. Photograph by the author.

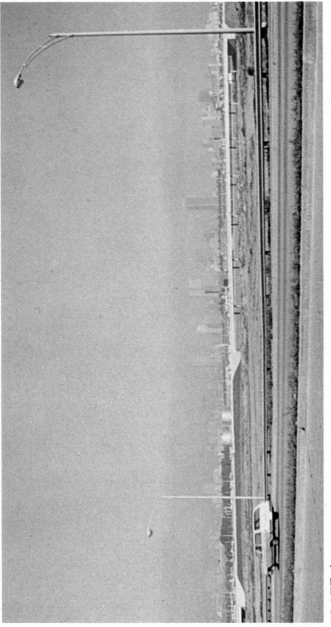

PLATE 1

PLATE 2

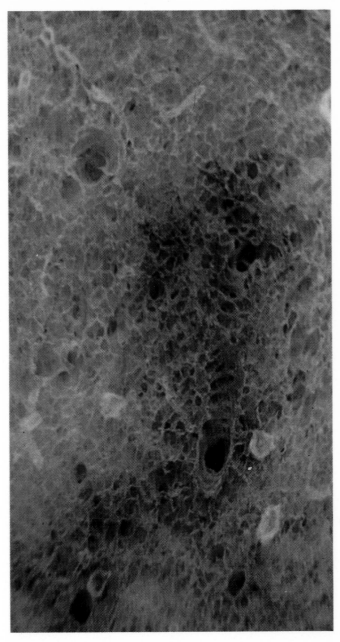

PLATE 3

PLATE 4

chapter one

THE CONTEMPORARY PROBLEM

A chronology of concern about air pollution may be extended almost indefinitely. Yet, in spite of the occasional voice of concern, there was little major action or study relating to air pollution until after the Second World War. A speaker in 1913 in Britain predicted that unless all coal mines were closed for a number of years the human race would be asphyxiated by the year 1950, and it is perhaps ironic that the first major smog episode in Britain occurred in London in 1952. Dr. Galton in 1882, in his inaugural address to the Sanitary Institute of Great Britain, spoke at considerable length about air pollution, mentioning not only smoke but the problem of horse manure. The determined efforts of the city of Pittsburgh in the U.S.A., during the nineteen thirties, to clean up its environment stood as a prototype for other cities in which the main problem was soot as a result of the burning of coal in open fires. Specific episodes of industrial pollution which occurred at Donora (Pennsylvania) and in the Meuse Valley in Belgium tended to be regarded as isolated incidents and not relevant to the general community of the modern city.

For seven days in December 1952, the city of London was shrouded by a major temperature inversion.[61] At that time the cheapest available method of domestic heating was the burning of coal in an open grate, and this generated a great deal of large particulate smoke and sulphur dioxide. The excess mortality over this period was approximately four thousand people, and it has been shown subsequently that similar heavy episodes of air pollution are associated with an increase over the usual running mortality, as noted later. This has been found now to be true in a number of different cities. In Britain the report of the Beaver Committee on Air Pollution (Cmd. 9322. London: H.M.S.O., 1954) concluded: "There can be no doubt that the effect of air pollution on health is wholly bad, whether measured positively in relation to growth, well-being, and joy of living, or negatively in terms of death, disease, and economic loss which goes with incapacity to work." In 1956 the Clean Air Act was passed in Britain, and this enabled a city to prohibit the open-burning of coal in designated areas. This has had a major effect on reducing particulate pollution and, to a lesser extent, sulphur dioxide levels in major cities in Britain (see chapter 6).

At about the same time, it was recognized in the U.S.A. that in the Los Angeles area there was a special air-pollution problem related to heavy automobile density, and to the special meteorologic conditions of that region. Although no major health disasters occurred, this was clearly a different kind of pollution from that existing in coal-burning cities, and obviously required detailed study and investigation.

As soon as inventories began to be made of air pollution, it became evident that the biosphere was being required to absorb and deal with staggering quantities of pollutants. In the New York and New Jersey area in 1966, for example, 231,000 tons of particles were released into the air during the year.[12] The amount of carbon monoxide generated by man has been formidable, and this gas at street level has reached undesirable concentrations in some cities.[9] A combination of circumstances enabled people for the first time to visualize the earth as a totality. Without doubt the photographs of the earth taken from the Apollo spacecraft brought home to people the fact that the earth's environment was not limitless, and we have come to learn that some of its internal balances are relatively delicate. The widespread and uncontrolled use of chemical fertilizers and pesticides, the problem of water conservation and protection, and the increasing evidence of deterioration of the air environment have all contributed to public requests for action if the progressive degradation of the environment is to be halted.

In the last fifteen years, a great deal of information has been acquired relating to air pollution. It is fair to say that we now have a good idea where most pollutants come from, that we are in a position to make provisional estimates of the comparative hazard of different pollutants, and that research has begun on the most economical methods of controlling their emission. Citizens of many countries have expressed a determination that future husbanding of our resources and protection of our environment must be at a much higher level than was true in the past. Sixty years ago smoking chimneys were synonymous with industrial progress, an increase in gross national product, and employment. Now we know that the costs of such a simple objective as an increase in gross national product, if accomplished with no regard whatsoever for the environmental consequences, will be too heavy for future generations to bear.

It has been argued that nothing short of a major social and political revolution will be required before the transition can be made from one kind of orientation of society to the other. Whether or not this proves true, it should be clear to everyone that a society which places the highest possible priority on the protection of its natural resources and the cleanliness of its air and water, and defends these against powerful industrial interests where necessary, would bear so little resemblance to the society in the western world of the past sixty years that the "revolution" would already have occurred within it.

The contemporary problem of air pollution is primarily a problem of cities. It is in cities that the major issues arise, and it is on their solution that the future of the city to a very large extent depends. Air pollution is of major concern to man and of only secondary concern in relation to birds, mammals, and ecology as a whole. It is true that very high levels of certain industrial effluents may produce damage to the environment, but the problems of high population density, transportation, and heating within the city environment pose the greatest difficulties for the future.

chapter two

THE NATURE
AND SOURCES
OF AIR
POLLUTANTS

SULPHUR DIOXIDE (SO_2)

Sulphur dioxide is a gas produced when any fuel containing sulphur is burned in the presence of oxygen. Sulphur is commonly present in coal (though to varying degrees) and in oil, and sulphur dioxide production is therefore a feature of the burning of both of these fossil fuels. Sulphur dioxide is very soluble in water and under special circumstances may exist in association with water, not only as SO_2 but also in the form of sulphuric acid (H_2SO_4) and as sulphate salts in suspended particulate matter. Under certain meteorological conditions, a sulphuric acid mist may be formed in the air from sources of sulphur dioxide.[14]

In table 1 are shown some main sources of sulphur dioxide calculated in 1966 for the entire United States,[14] and as estimated for the city of Montreal in the same year. For the United States as a whole, the burning of coal still represents the major source of SO_2, much of this coal being used for electricity

TABLE 1

SOURCES OF SULPHUR DIOXIDE

Source	Calculated Annual Tonnage for USA (1966)	% of Total	Estimated Annual Tonnage for Montreal (1966)	% of Total
Burning of coal	16.6 million	58.2	30,600	11
Burning of oil	5.6 million	19.6	162,000	58
Smelting of ores	3.5 million	12.2	0	0
Refinery operations	1.6 million	5.5	73,440	27
Refuse incineration	0.1 million	0.4	360	—
Miscellaneous	1.2 million	4.1	12,600	4
Total Emission	28.5 million	100.0	279,000	100

Data for the United States are taken from reference 14 and data for Montreal from references 23 and 67. The figures should be regarded as only approximate in both sets of data, but they give an indication of proportional sources of sulphur dioxide. More recent data from Montreal have indicated that the annual tonnage of SO_2 being produced is considerably greater than indicated here, and in winter the output from domestic sources was computed to be 1,390 tons per day, and from industrial sources 480 tons per day, for a total of 1,870 tons per day (see reference 35).

generation. In Montreal, domestic heating with oil and refinery operations constitute the major sources. Sulphur dioxide is produced in substantial tonnage in some ore operations, copper smelting in particular being associated with considerable emission of the gas. Between 1926 and 1930, the emission of SO_2 from the stacks of the large lead–zinc smelter at Trail, B.C., attained high levels, with a maximum of about 20,000 tons of SO_2 per month.[68] This resulted in widespread damage in Canada and in the adjacent part of the United States, but within the following decade a large industry was created to convert the sulphur-containing gases to sulphuric acid, ammonium sulphate, and fertilizers, and a better than 91 per cent recovery rate of sulphur dioxide has been achieved.

The emission of sulphur dioxide from the nickel plant at Sudbury, Ontario, has also been the subject of intensive study, and maximum half-hour concentrations ranged from 0.63 ppm (parts per million) at a distance of forty miles from the plant to 3.64 ppm sixteen miles downwind. Injury to agricultural and forest species of vegetation occurred when concentrations reached 0.95 ppm for one hour, or 0.55 ppm for two hours.[36, 68] The emission of major quantities of sulphur dioxide from the industrial regions of Britain and of the Ruhr in Germany is believed to have resulted in detectable levels over parts of Scandinavia, and possibly in alteration of the acidity of rivers many hundreds of miles from the sources of the sulphur dioxide.

Within the city environment, the burning of coal or of oil represents the two major sources of sulphur dioxide. Coal may be burned to generate electricity, and oil is commonly used for home heating in North America, Scandinavia, and other parts of the Northern Hemisphere. Inevitably, recovery of sulphur dioxide is impossible from multiple domestic sources, and control depends on limiting the sulphur content of fuel which may be sold in a given region (see chapter 6).

Since sulphur dioxide is reasonably simple to measure, there is now a very considerable amount of data from many different cities in the world describing the level of sulphur dioxide reached on a day-to-day and month-to-month basis. The data from different cities show such general uniformity of pattern that there is no reason to suppose that individual estimates reported from different centres contain serious errors. It is a remarkable feature of sulphur dioxide pollution that the levels in any individual city when plotted as a function of time show an almost linear relationship on a logarithmic scale. This method of plotting sulphur dioxide concentrations is very useful, since by it one can predict a probable level existing for one period of time on a basis of data furnished on a longer or shorter term. Figure 1 shows some of the data for different cities plotted in this way, and the generally parallel nature of the lines for individual cities will immediately be evident. It follows from this relationship that, although it is not known whether high peak concentrations are more potentially harmful than lower-level exposures for longer terms, this is largely an academic question since any city which significantly reduces its sulphur dioxide pollution would experience a predictable change both in peak-level concentrations and in averages over a period of a month or a year.

Figure 6 (p. 53) shows the sulphur dioxide levels which existed during the London fog episode of December 1952, and it can be seen that levels above 0.25 ppm occurred continuously for approximately five consecutive days in that month. In addition, these were associated with very high smoke concentrations, as noted later.

In cities such as Montreal or Toronto, variations in the level of sulphur dioxide in the atmosphere from one part of the city to another are considerable. In Montreal, for example, an approximately fourfold variation in mean yearly average exists between the station showing the highest mean average sulphur dioxide over a year (of 0.08 ppm) and a station four miles away in a different part of the city which showed only a quarter of this level.[67] The actual concentration existing at any place is a complex function of the emission rate and the meteorological conditions. In Montreal, in winter, with heavy snow cover, it is common to have a temperature inversion between 2:00 A.M. and 7:00 A.M. In this situation, a cold layer of air at ground level and extending a few hundred feet above it becomes trapped beneath a zone of warmer air. Since this period coincides with the lowest temperature of the winter night, when a great deal of oil is being burned, sulphur at street level not

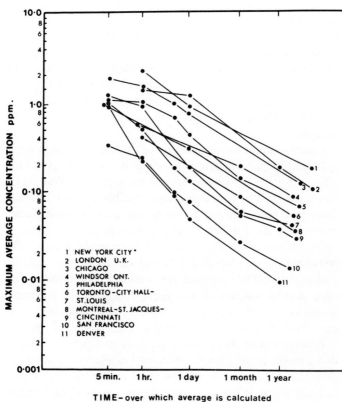

FIGURE 1.   SULPHUR DIOXIDE LEVELS IN DIFFERENT CITIES

Data from reference 14 with additional points for Toronto and Montreal from references 67, 72, and 73. Notice that, whichever city is considered, there is a generally constant relationship between the peak concentrations of sulphur dioxide observed over a short time interval and the annual average concentrations. Thus if legislation leads to a general reduction in the amount of sulphur dioxide emitted, both the annual average value and the peak values observed during a year would be expected to fall. Most of the data on this chart relate to the period between 1965 and 1969.

infrequently reaches a maximum during the early hours of the morning (see plate 1). A wind arising shortly after 7:00 A.M. commonly leads to some dispersal, but if the meteorological conditions are stable for two or three consecutive days the sulphur dioxide level during that period of the year quickly rises.

By sampling the air over Montreal and at street level, it has been shown that the regions surrounding the island of Montreal contribute little to the concentration of SO_2.[35]

As noted later, the fact that sulphur dioxide has been very extensively measured has meant that there has been a natural tendency to assume that effects being observed in the population are closely related to sulphur dioxide levels. As explained in chapter 4, such a conclusion must be interpreted with caution.

Calculations of levels of sulphur dioxide coming from single industrial sources are complex. As has been noted, several observations have been made around the smelter in Trail, B.C., during the 1930s, and also around Sudbury. Other observations have been made of power plants, and one such study[14] determined the sulphur dioxide concentration in a ring, of radius three to four miles, centred on a power station. The power station was burning 430 tons of 1.5 per cent sulphur coal per hour. The maximum three-minute concentration during a year of observation was about 0.62 ppm, the maximum hourly average was about 0.47 ppm, the maximum daily average was 0.11 ppm, and the annual average was 0.027 ppm.

Comparison of these data with those in figure 1 indicates that a single major source of this kind can produce, in a rural environment, approximately the same concentrations as exist in a city where the sources of emission are multiple. Presumably people living within the critical area of such a plant might be exposed to greater levels of sulphur dioxide than the city dweller. As will be noted later, control methods have been successfully applied to the problem of sulphur dioxide pollution. Particular success has been recorded in London, England, and in the city of New York (see figures 8 and 9), and it seems clear that episodes of high SO_2 pollution such as have occurred in those cities in the past will become progressively less frequent in the future.

Emission of sulphur dioxide from industrial sources can also be easily controlled, and thirty years ago the recapturing of sulphur from factory effluents could have been economic. But with the present excess of sulphur in the world it is of little value to a company to recapture the element. Restrictions on the sulphur content of fuel oil that may be burnt in the city have meant that refineries have had to develop ways of removing it from high sulphur crude oil, and of recapturing it during the removal process. In the absence of a market for sulphur, such processes inevitably lead to a higher cost of fuel oil.

PARTICULATE POLLUTION

The modern city produces a large number of particles which are variable in size and in composition. Relatively large particles (larger than 10 microns) create visible smoke and are produced by burning coal or by incineration processes from poorly controlled appliances. These particles, being large, tend to settle to the ground relatively quickly and hence constitute an easily measured dustfall. Before 1950, this dustfall was commonly measured in units such as tons per square mile per month, and in many cities this figure exceeded 70 tons. As will be noted later, such large particles are trapped in the human nose and do not penetrate far into the human lung. In most cities, the burning of coal in open hearths has greatly diminished, either as a result of legislation (as in the Clean Air Act of the United Kingdom) or for economic reasons. Coal-burning railway locomotives have ceased to be economical and have disappeared from the scene in North America, and the decline in gross dustfall which has been recorded in Montreal since 1950, and in other cities, is largely attributable to the dwindling of this sort of air pollution.

Oil burning produces smaller particles, and removal of large particles from incineration processes has also contributed to a dramatic decrease in particle size in many cities over the past ten years. Cars and trucks produce small particles, as are those generated by industrial sources of pollution, particularly such plants as cement works, iron foundries, and metal-processing mills of different kinds. Particles may also be contributed to the air of the city from brake linings of vehicles and from a wide range of urban activities, including construction work, road repairs, and rubbish incineration.

In table 2 is shown a recent inventory of the sources of

TABLE 2

SOURCES OF PARTICULATE EMISSION FOR THE
CITY OF NEW YORK, NOVEMBER 1969

SOURCE	QUANTITY TONS/YEAR	% OF TOTAL
Space heating	22,300	32.3
Municipal incineration	13,330	19.3
On-site incineration	12,690	18.4
Mobile sources	9,900	14.3
Power generation	6,400	9.2
Industrial	4,500	6.5
	69,120 tons/yr	100.0

These data are taken from reference 37. Mobile sources refers principally to city traffic.

particulate emission for New York City, and table 3 provides a similar calculation for the city of Montreal. In New York, space heating is believed to be the major contributor of particulate pollution.[37] Since the dustfall index no longer reflects the predominant particle pollution, particulates in the air are commonly measured either in terms of a coefficient of haze (coh unit) or in units of micrograms per cubic metre ($\mu g/m^3$). Table 4 illustrates some particulate levels for a number of United States and Canadian cities. In the modern city, the average particle size expressed as the mass median diameter varies between four and eight microns. There are, however, many particles of a much smaller size range, though their mass contribution to the total may be small.[12]

When one considers the highly varied sources of particles, it is hardly surprising that their composition is extremely complex. Randomly sampled particles contain, as one would expect, common materials such as carbon and silica, but also a very wide range of metals such as cadmium, chromium, copper, iron, lead (small lead particles are produced by automobiles using leaded gasoline), manganese, nickel, and many others.[12] In addition to these, asbestos particles have been found free-floating in city air, and these are believed to come mainly from construction work and possibly also from the

TABLE 3
ESTIMATED POLLUTANT EMISSIONS IN MONTREAL IN 1966 IN TONS PER DAY

	VEHICLES		OIL		HEAT AND POWER		GARBAGE INCINERATION		OIL REFINERIES	OTHER INDUSTRIAL	TOTAL
Daily Consumption	GASOLINE 2,340,000 gallons	DIESEL 530,000 gallons	Light 1,700,000 gallons	Heavy 2,300,000 gallons	COAL 2,100 tons	GAS 103 million cubic feet	MUNICIPAL 1,000 tons	DOMESTIC 225 tons	BBl 340,000	[No detailed estimates]	
Particulate	16	30	9	12	103	1	5	3	27		206
Sulphur oxides	14	12	75	383	85	1	1	0	204		775
Nitrogen oxides	106	58	62	143	30	9	1	0	20		429
Hydrocarbons	535	84	4	6	21	4	10	1	210		275
Organic acids	5	8	13	18	31	3	10	0	3		92
Aldehydes	12	4	2	1	2	1	1	0	3		26
Ammonia	2	1	1	1	3	0	1	0	3		12
Carbon monoxide	3,400	15	2	—	29	—	0	3	0		3,449
Total	4,090	212	168	564	304	19	29	8	470	150	6,014

This estimate was compiled by the City Health Department and represented the best possible appraisal, at that time, of sources of pollution in the city of Montreal. More recent data indicate that the sulphur dioxide output from the oil refineries is considerably greater than the estimate given in this table, but the general subdivisions of pollutant contribution are probably valid. A similar computation for 1961, based on daily consumption figures, revealed that almost all pollutants had increased by between 30 and 50 per cent over the five-year period between 1961 and 1966. The expression of all pollutants in terms of tonnage is of some use for purposes of comparison, but does not necessarily denote a proportionate importance. More recent data for SO_2 from Montreal indicated that in winter domestic sources contribute 1,390 tons per day, and industrial sources 480 tons per day, for a total of 1,870 tons per day (see reference 35).

TABLE 4

AVERAGE PARTICULATE CONCENTRATION ($\mu g/m^3$) IN 401 U.S. CITIES 1957–1967

POPULATION	<60	60–120	120–160	160–200	>200	NUMBER OF CITIES
>3 million	—	—	1	1	—	2
1–3 million	—	—	2	1	—	3
700,000–1 million	—	3	4	—	—	7
400,000–700,000	—	9	7	2	—	18
100,000–400,000	3	61	29	5	1	99
50,000–100,000	2	64	18	6	3	93
25,000–50,000	5	48	12	3	3	71
10,000–25,000	7	46	7	4	—	64
<10,000	6	33	3	2	—	44
						401

These data are taken from reference 12. Notice the wide variation in particulate pollution that may be encountered in smaller cities of comparable size. The data in this table should be compared with those in table 5.

brake linings of automobiles, though their source to this point has not been precisely identified.

In individual cities of a given population size there is considerable variation in average particle pollution, as may be seen from table 4. For example, for cities in the population class of 100,000 people, the average particle concentration may vary from 60 $\mu g/m^3$ to over 200 $\mu g/m^3$ on an annual average basis. Such differences usually are most closely related to the quantity of fuel being burnt for domestic heating and to the presence of particle-producing industrial processes within the city area, or close to it.

In such cities as New York, Montreal, and Toronto, the fine particulate pollution in winter causes deposition of a thin black film on the inner side of the car windshield when the defroster is in use. These particles may be observed by wiping the inside of the windshield with a paper tissue; the deposited film may be easily seen after a matter of only a few days of city driving. Table 5 shows some comparative particulate-pollution data for U.S. and Canadian cities. Too much precision should not be attributed to these numbers, but they do provide useful general

TABLE 5

PARTICULATE POLLUTION LEVELS IN SOME U.S. AND CANADIAN CITIES

CITY	SAMPLING SITE	OBSERVATION DATES	MEAN μg/m³	MAXIMUM OBSERVED
Canada				
Windsor	City Hall	1969	177(A)	563
Hamilton	City Hall	"	168(A)	786
Toronto	College Street	"	138(A)	482
Ottawa	Kenson Bldg.	"	96(A)	159
Orillia	Centre	"	88(A)	290
Kingston	City Hall	"	81(A)	228
Montreal	Drummond St.	1968/69(64)	136(A)	437
Montreal	Blvd. St. Michel	" (22)	318(A)	684
Montreal	Notre-Dame East	" (40)	104(A)	206
U.S.A.				
Chicago	Averaged	1961–65	177(G)	
Philadelphia	"	"	170(G)	
New York	"	"	135(G)	
Buffalo	"	"	126(G)	
San Francisco	"	"	80(G)	
Miami	"	"	58(G)	

These data have been taken from multiple sources in the bibliography. (A) refers to the arithmetic mean of many observations and, in Montreal, is the mean of a number of paired observations as indicated by the figures in parentheses after the observation date. (G) after the American data indicates that the geometric mean has been used. In actual practice, this is not greatly different from the arithmetic mean and for general purposes the two figures may be compared. Notice from the Canadian data that the maximum level observed bears a general correspondence to the average level over a much longer period, and permits the generalization that the maximum level observed at any one sampling site is between four and five times the mean averaged over a long period. These levels should be compared with the proposed Canadian Air Quality levels (see Appendix).

information for comparative purposes. It is interesting to note that the particulate pollution in London in the disaster of 1952 reached 1500 μg/m³ (see figure 6). In prairie cities of Canada such as Winnipeg, particulate pollution is very low in winter since natural gas is burned for heating, but dust blown off the land in summer results in high particulate counts during some months.[33, 76]

During the past few years, many refinements have been introduced into the difficult problem of measuring particle concentration. Particles from 0.1 to 10 microns comprise the bulk of the particulate mass, and a major fraction of the numbers in city air. These may be measured by simple filtration samplers in which dust collected on the filter may be assessed by measuring the incident light transmitted through the paper as compared with the control. More refined measurements are needed for particles smaller than 0.1 micron, and electron microscopes have been used for identification and to make estimates of the numbers of those that are extremely small in dimension.[12]

As will be noted later, particles are an important constituent of the air-pollution problem of the modern city. Not only do they play a predominant role in reducing visibility (see chapter 3) but they also contribute in a major way to the economic costs of air pollution as a consequence of the soiling of clothes and buildings. The exact role played by particles in producing effects on health is far from clear (see chapter 4), but at this point it may be noted that there are several reasons for contemporary concern relating to particulate pollution.

The first of these is the fact that the average size of particles in city air is now smaller, because the larger particles have been eliminated by control methods and by a considerable reduction in coal burning. These smaller particles are more likely to be retained in the human lung. Figure 4 shows the deposition rate of particles of different size in the human lung, and it will be noted that when they are much smaller than 4 microns the rate of deposition in the human lung increases sharply. It was a common experience of those living in London during and after World War II that on a day of fog, if one blew one's nose, one observed the black material which had been trapped in it. Particles one-tenth the size of these are carried into the lung, and there is reason to be concerned with the effects these may have.

Secondly, it must be remembered that particle pollution in the city occurs in the presence of gaseous pollution, usually either sulphur dioxide or oxides of nitrogen. There is some reason to suppose that the effects of these gases may be potentiated in the lung if the gas is adsorbed onto a particle small enough to be deposited within the lung. Whether or not this is true, there is reason to be concerned about the combination of such particles (small enough to be deposited in the lung) with these and other gases in the city environment.

It may be noted from table 3 that although isolated instances of emission of black smoke from apartment blocks or other buildings are quick to be noted by the general public, and are often reported to the air-pollution control authority, this source of air pollution is in general quite small. Furthermore, visible black smoke of this kind consists of larger particles; although the inconvenience of visible soot deposition on a line of laundry is unquestionable, in the longer term there are good reasons for being more concerned with the small-particle pollution which has become a feature of city living.

In addition, many complex industrial processes give rise to particulate pollution, and in some instances these particles have a high metallic content. As will be noted later, iron is relatively inert in the human lung, but the same cannot be said for particles containing cadmium, manganese, or nickel. Although the small lead particles from gasoline exhaust are deposited in the lung, the total dose acquired by this means would be unlikely to produce significant clinical evidence of lead poisoning; but children being brought up adjacent to major highways may certainly acquire from this source a load of lead higher than desirable. As will be noted later, for many years asbestos particles have been found in the lungs of city dwellers, but the likelihood that these play much part in generating lung disease seems remote (see chapter 4).

POLLUTANTS FROM AUTOMOBILES

The automobile is such an important cause of air pollution that it merits separate description. For reasons that will become clear, the gasoline engine is not only an important source of air pollutants, but the density of traffic and the number of automobiles being operated in the city environment mean that the total production of pollutants from this source in a city area is very considerable. Different pollutants cannot be compared

with each other on a basis of tonnage, and there is little meaning in equating tons of a gas, such as carbon monoxide, with
tons of particles less than a micron in size, or even with other
gases. Furthermore, although the automobile has become a
pollution problem in the city environment, there are reasons
for stating that from the long-term cross-country haulage point
of view there really are few objections to the gasoline engine
(see chapter 5). Uncritical denunciations of the automobile for
all purposes are unhelpful, as is an unwillingness to concede
that congestion of contemporary vehicles in a city with high
sunlight intensity and relatively little air movement is justifi-
ably a cause of concern.

Carbon Monoxide (CO)

Carbon monoxide is the most commonly occurring air pollutant, and if measured in tonnage the total emissions of this
gas to the atmosphere exceed those of all other contaminants
added together. An estimate from the United States for 1968
puts the total emission of carbon monoxide at more than 100
million tons per year.[9] Sixty-three per cent of this total is believed to be contributed by transportation, and a major part of
the percentage comes from gasoline-powered engines. Industrial processes (metal industry, gasoline refineries, some paper-
processing) are believed to contribute about 11 per cent of the
total CO, and miscellaneous sources (including such things as
forest fires) contribute about 16 per cent of the total. Carbon
monoxide is generated by any process of incomplete combustion, and it is an important constituent of cigarette smoke from
which it is absorbed into the body if the smoke is inhaled into
the lung (see chapter 4).

There are several important factors, based on the design of
the engine, which determine the emission rate of carbon
monoxide by motor vehicles but, regardless of that factor, the
emission of carbon monoxide on the basis of weight per mile
travelled is proportionately much less at speeds over thirty
miles an hour than at average route speeds of less than 10
miles an hour. Thus it has been computed[9] that an increase in
average urban route speed from 25 to 35 miles an hour can
result in a reduction of roughly one-fourth in the mass of CO

emitted to the atmosphere. The amount of carbon monoxide given off by diesel vehicles is very much less than by gasoline vehicles, largely because of the difference between the air–fuel ratio in these two types of engines (see table 3). Major reductions in carbon monoxide emissions from gasoline engines are occurring as a result of redesign and stricter specifications, and the total carbon monoxide discharges from motor vehicles are expected to fall continuously as more and more vehicles on the road conform to these new standards.[63]

As would be expected, the levels of carbon monoxide occurring in a city are directly dependent on the meteorological conditions and the traffic density. Peak concentrations of as high as 230 ppm have been measured at street level on heavy traffic days in London. In 1967, 235 samples out of 15,000 collected from over three-hundred locations in Paris exceeded 100 ppm of CO. More usual measurements are of the order of between 5 and 40 ppm of CO, the maximum concentration recorded in Toronto being 29 ppm over a number of sampling stations,[72, 73] and a reading of 40 ppm being recorded on the Decarie Expressway over one holiday period in Montreal.[23] The annual average level for Montreal as a whole is between 5 and 10 ppm.

Under special high density traffic conditions, and particularly during traffic jams, levels of this gas may rise precipitously. Apart from such abnormal conditions, however, the general level of carbon monoxide existing in the city environment produces a change in carbon monoxide in the blood of less than 1 per cent. This is to be compared with the levels of up to 10 per cent found in the blood of heavy cigarette smokers, and it is clear from present data that it would be virtually impossible in the ordinary urban environment for traffic to produce levels of carbon monoxide hemoglobin in the blood more than about a tenth of those produced by heavy cigarette smoking. This point is discussed in detail in chapter 4.

Oxides of Nitrogen

Any combustion process occurring under pressure and with heat is likely to result in the fixation of nitrogen and oxygen which are normal constituents of air. The oxide of nitrogen initially produced is usually in the form of NO which becomes oxidized in the atmosphere relatively quickly to become the more irritant NO_2. This reaction is discussed later. The total tonnage of these oxides emitted in the United States in 1968 was estimated to be approximately 20 million tons per year.[11] About 35 per cent of this was believed to be produced by motor vehicles, and a large percentage (48.5 per cent) was caused by fuel combustion, particularly coal and natural gas. Miscellaneous sources contributed the balance.

The annual nation-wide emissions of nitrogen oxides in the United States have increased from an estimate of 16.7 million tons in 1966 to 20.6 million tons in 1968, and this figure is predicted to continue to rise.[63] Oxides of nitrogen measured in city air and expressed in ppm reach monthly averages as high as 0.2 in Los Angeles, but for a city such as Chicago they vary between 0.04 and 0.14 ppm, depending on month of the year. Peak concentrations do not exceed approximately 0.5 ppm.[11] For the United States as a whole, 10 per cent of cities with populations less than 50,000 show a yearly average equal to or exceeding 0.06 ppm. In the population range from 50,000 to 500,000, half of the U.S. cities equal or exceed a yearly average of 0.06 ppm, and in those with a population of over half-a-million 85 per cent of cities equal or exceed 0.06 ppm of oxides of nitrogen on a yearly average.[11]

From Toronto it has been reported that levels of nitrogen oxides exceeded 0.2 ppm only during 6 of the 7000 hours of recording in downtown Toronto, and a daily average of more than 0.1 ppm was recorded on only one day of the period of observations.[72, 73] Data on this pollutant from other Canadian cities are not yet generally available.

These very low levels of oxides of nitrogen might seem to be of no consequence, but they are important, not only because there is some evidence (as noted in chapter 4) that levels in the range of 0.06 ppm as an annual average may be associated with increased respiratory disease, but also because oxides of nitrogen form the basic material from which the chain reactions known as photochemical smog begin (see below).

Hydrocarbons

Hydrocarbons are produced primarily by automobiles. The principal compound is methane, which may average 3 ppm in urban air; also present are other alkanes and alkenes, cephalene, and aromatic hydrocarbons such as benzene and toluene.[10] The highest concentration of benzene measured in Los Angeles air over a 26-day period was 0.057 ppm, and most of the other aromatic hydrocarbons are present in lower concentrations than this.[10] Much of the smell of diesel and gasoline exhaust is produced by the presence of aldehydes, and formaldehyde, which is one of these, reaches concentrations of 0.05 ppm in the air of Los Angeles.

Small though these concentrations are, they are significant in terms of affecting plant life at very low concentrations, as noted in chapter 3, and they are also important in relation to the reactions that occur between gasoline exhaust and sunlight, as noted below. There is a close relationship between hydrocarbon effluent and automobile density, and the fluctuations in air level follow closely the fluctuations that may be observed in carbon monoxide level.

As a consequence of some industrial processes, and also from gasoline and diesel vehicles, a number of polycyclic hydrocarbons have been identified in the atmosphere of cities. Particular attention has been given to benzpyrene compounds since these are present in cigarette smoke and in high concentrations are believed to be capable, under some circumstances, of inducing cancer. Concentrations in city air are generally very low, and the measurements are technically very difficult to make. Background values in the rural environment are approximately 0.5 μg/1000 m^3. In Vienna, during the winter months, levels of 123–165 μg/1000 m^3 have been recorded, and high levels may exist in such environments as diesel bus garages or in the presence of special industrial processes. It has been tempting to associate the presence of benzpyrene in city air with lung cancer, particularly since the same compounds are present in cigarette smoke, but careful studies of the incidence of lung cancer in men exposed to high levels of these compounds have so far failed to reveal any close association between the incidence of cancer of the lung and the level of exposure.

Particulates

The most important single particulate emitted from the automobile is lead. There has recently been considerable concern over this aspect of pollution, though the removal of lead from gasoline has been largely occasioned by its undesirable effects on catalytic systems designed to reduce the emission by automobiles of hydrocarbons and oxides of nitrogen.[17] Levels of lead in the atmosphere as a consequence of automobile traffic vary between 0.1 and 5.0 μg/m^3.[24] An average weekly lead concentration of as much as 8 μg/m^3 has been observed. As would be expected, the variation is considerable, even within the same general area, but temperature inversions may produce high concentrations of atmospheric lead if they coincide with high traffic density.

The emission of lead is less important in the short-term than in the long-term problems it will create; evidence has shown that once given off into the atmosphere it remains on vegetation for a long time. Furthermore, until recently, it had not been generally appreciated how much lead was being added to gasoline. Fifty million pounds of lead were discharged into the ambient atmosphere within the State of California each year, a not inconsiderable total; but it was only 10 per cent of the total amount of lead being combusted in the United States through gasolines. For the United States as a whole, approximately 180,000 tons of lead are put into the air annually from motor vehicles. It has been shown conclusively that the lead in the atmosphere in San Diego and vicinity is the same compound that is added to gasoline,[24] and, although the immediate risks are slight, the long-term consequences of this pollutant, as noted in chapter 5, have provided sufficient reason for legislative control.

Photochemical Smog

When gasoline exhaust containing oxides of nitrogen and hydrocarbons is discharged in quantity into the air in the presence of intense sunlight and under still wind conditions, there occur a series of reactions by which secondary products are formed, as depicted in figure 2. In addition to the production of ozone, the reaction leads to the formation of unstable and highly irritant compounds of the general nature of peroxyacetyl nitrite, commonly referred to as PAN compounds. They are believed to be responsible for the irritation of the eyes caused by the products of these reactions. It has recently been shown that the process of formation of ozone and of these compounds is accelerated by the concomitant presence of carbon monoxide in the air, by a presumably catalytic process that is not well understood.[78] Figure 3 indicates the sequence in development of concentrations of NO_2 and ozone, on a day of bright sunlight and high air stability in Los Angeles. It will be

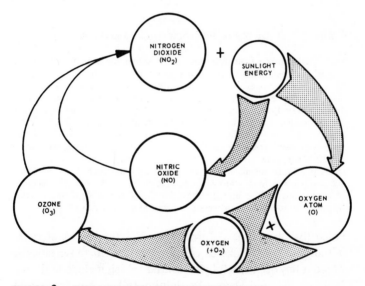

FIGURE 2.   FORMATION OF PHOTOCHEMICAL SMOG

This diagram, taken from reference 13, shows the sequence of events occurring as ozone is formed from nitrogen dioxide and nitric oxide in the atmosphere. It is believed that the presence of hydrocarbons accelerates this reaction, and it has recently been shown that the presence of carbon monoxide has the same effect. This is an important observation since carbon monoxide is likely to be present at the same time as oxides of nitrogen because both are derived from automobiles.

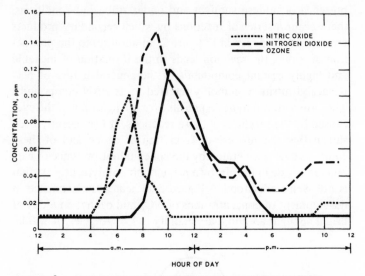

FIGURE 3.   SEQUENCE OF PHOTOCHEMICAL SMOG FORMATION

This chart is taken from reference 13. It shows the sequence of formation of nitrogen dioxide and ozone from nitric oxide in Los Angeles on a day of relatively still air and bright sunshine in July 1965. The first gas to appear is nitric oxide which reaches a peak at about 7 A.M. during the early morning rush hour. This is rapidly oxidized to nitrogen dioxide which reaches its highest concentration at about 9 A.M. From this gas, in the presence of bright sunlight, is formed ozone which reaches its peak concentration at about 11 A.M. All of these substances steadily disappear so that by 4 P.M. they are present in only small amounts. More recent work has shown that the highest ozone levels are reached not in downtown Los Angeles but in the suburbs, because of slow wind drift inland. In these localities, the peak ozone concentration is almost always observed to occur at about 1 P.M.

noted that the ozone is formed three or four hours after the initial emission of oxides of nitrogen, and in those cities in which it has been carefully measured, which include St. Louis, Philadelphia, and Denver in addition to Los Angeles, a peak concentration is normally recorded at approximately 12 noon.[13]

Of particular interest has been the finding that the slow drift of wind inland in Los Angeles means that the peak ozone concentrations occur in the suburbs of that city rather than in the immediate downtown area.[62] One would predict that the oxidant pollution produced by this mechanism would be maximal in cities with very high automobile density, and where the sunlight is bright. Levels in the Los Angeles region have been measured to exceed 0.5 ppm for short periods of time, and levels of up to 0.3 ppm have been recorded in cities on the eastern seaboard of the United States.[13] Very few measurements are available for Canada, but a level of 0.1 ppm of oxidant concentration was exceeded during only 36 of over 6000 hours of surveillance in downtown Toronto. The maximum reading in Toronto in 1968 was 0.18 ppm.[73]

Photochemical smog might be thought of as a problem only in highly developed countries, and in cities such as Los Angeles. However, if the automobile density is high enough, and the sunlight intensity great enough, the same reactions would be expected to occur anywhere. I have recently been sent photographs taken at 7:00 A.M. and at 11:30 A.M. on a clear day of high sunlight intensity in Teheran. The automobile density in the city is approximately the same as in Montreal, and the photochemical smog obscuring the view of distant hills was very evident in the photographs. It must be presumed, therefore, that any city with high automobile density and relatively stable air conditions may experience this sequence of changes, and one would expect Mexico City, Buenos Aires, or Bombay to experience the same phenomena on certain days of the year. Unfortunately, so far there has not been an international effort to make comparable observations of oxidant level in such cities, but no doubt these will be forthcoming.

Data from northern Europe on oxidant pollution up to 1970 indicated that the sunlight intensity was not often great enough to produce levels similar to those observed in the United States. Recently (October 1971), however, a severe photochemical smog episode was reported from Rotterdam and its vicinity, and an account from Britain stated that photochemical air pollution was more important than was previously thought. No precise data have yet been made generally available. As noted below, the oxidants produced in photochemical smog have considerable effect on neighbouring vegetation, and certainly produce noticeable physical discomfort amongst the population.

SPECIAL POLLUTANTS

Hydrogen Sulphide

Hydrogen sulphide may be emitted from Kraft paper mills and from some refinery processes. The gas is likely to be a nuisance only to those living in proximity to one or other of these processes, but its odour is detectable at levels of 0.5 ppm. It is not known to have any adverse effects until concentrations of 100 ppm are reached, and therefore the gas constitutes a nuisance by virtue of its disagreeable smell rather than its hazard. Recently developed techniques have done much to diminish the emission of this particular pollutant.

Hydrogen Fluoride

Hydrogen fluoride is likely to be discharged during phosphate fertilizer production. High concentrations emitted over a period of years may result in undesirably high levels of fluorine in vegetation around the perimeter of the factory, and a hazard by virtue of the fact that such vegetation may be eaten by animals and later passed on to humans. For this reason the fluoride emissions from such plants have been carefully monitored; although such levels have in the past been high enough to produce evidence of fluorosis in cattle, human effects have not been proved. Nevertheless, hydrogen fluoride is an important air pollutant and obviously has to be carefully monitored if it is known to be present.

Toluene and Paint Solvents

In the Los Angeles region there has long been legislation defining the types of paint solvents which may be used in that area. In the presence of high intensity sunlight, considerable evaporation of these mixtures occurs as the paint is used, and may contribute a significant load of hydrocarbons and other compounds to the atmosphere. Similarly, toluene and other solvents may be emitted from factories producing paints and similar commodities, though these would rarely be in high enough concentration to produce undesirably high levels in the community as a whole. However, it must be remembered that a particular apartment block situated close to such a plant may be experiencing much higher levels of such effluents than the city as a whole; because a wide variety of such solvents are now being used in paint, supervision and control of their release has been thought to be necessary in a number of countries.

High Density Particulate Pollution

From some industrial processes, particularly those involving metallurgical operations, there may be a very high particulate effluent. The process of fusing iron and silica in an electric furnace gives rise to a very sizable particulate effluent, and from one group of such factories upwind from Montreal the average output in a 24-hour period of particles less than 1 micron in size is in excess of 25 tons (see plate 4). Such massive pollution by small particles is unusual, however, and in many countries the maximal particulate effluent permitted from various industrial processes is closely controlled. It must be remembered, though, that such metallurgical processes may be generating particles with a high content of some particular metal, as for example manganese, chromium, cadmium, or beryllium. The hazards of beryllium as an environmental pollutant have been recognized for some years, and the necessity for very strict control of its emission in a particulate form has been generally accepted.

All of these special pollutants are produced in association with specific industries, and may be particular hazards to small populations living very close to the industrial process concerned. None of them is likely to become of general importance as an urban pollutant, but the capability of an individual factory to subject a population to undesirably high levels of these and other substances must always be remembered.

INDICES OF POLLUTION

A number of indices combining measurements of different pollutants have been proposed, and some of these are in use in different cities.

A composite index devised for the Bay area of San Francisco was called the "combined pollutant index" (CPI) and was calculated as follows. Oxidants and nitrogen being measured in concentration of parts per 100 million, and carbon monoxide in ppm, and the coefficient of haze (COH), being directly measured, the index was calculated from the following equation:

$$CPI = 2 (OX) + (NO_2) + (CO) + 10 (COH).$$

Such an index deals particularly with photochemical pollution and does not take account, for example, of sulphur dioxide.

An index for New York City using sulphur dioxide, carbon monoxide, and the haze coefficient has also been designed; one devised by M. H. Green, for air-pollution measurements in Sarnia, Ontario, made use of sulphur dioxide and the coefficient of haze. By calculating an SO_2 index and a COH index the two were combined to give an air-pollution index (API) which was in fact the mean of the two individual indices so calculated. Other air-pollution indices have been proposed in different countries, and Ontario has made use of an index based on the 24-hour average concentrations of sulphur dioxide and of COH measurements.[65] This equation for the air-pollution index for Toronto is as follows:

$$API = 0.2 (30.5 COH + 126.0 SO_2)^{1.35}$$

The equation yields a number commonly lying between 30 and 100 and which indicates the combined levels of haze and sulphur dioxide. It is believed that adverse effects will not occur if the index is below 100, and the equation was partly devised to set a level of 100 as a threshold below which adverse effects would not be expected.

Whether preparation of this kind of index serves any useful purpose is doubtful. It would naturally be convenient if air pollution could be expressed in a single number, and there are many precedents in medicine for attempts to classify severity by some arbitrary addition of a few of the many variables involved. One of the disadvantages of such indices is that, given a single number, it is not possible to calculate what the individual readings of the constituents were at any given time, and any careful analysis of the air pollution in a particular region should deal with the presence of different pollutants individually in the first place. If sulphur dioxide is to be used as an index of the ambient air-pollution level it is better expressed in ppm on an individual daily basis, and the particulate pollution should be independently expressed. Neither of these may bear much relationship to the level of oxidant pollution, and in other regions the levels of oxides of nitrogen and ozone in the atmosphere will be of more general importance than either particulate pollution or sulphur dioxide. Whereas there can be little objection to calculating some composite index, it is essential that the data from which it is computed are simultaneously made public, since an index as a single number is by itself of very little use.

chapter three

GENERAL EFFECTS OF POLLUTANTS

EFFECTS ON PLANTS AND VEGETATION

It is remarkable that different plants show greatly varying sensitivity to specific pollutants. Some plants are so sensitive to very low levels of fluorides or oxidants that they may even be used to detect the presence of minute quantities of these substances. Others are highly resistant to these and other pollutants. In table 6 a few of the more sensitive plants are listed, together with the approximate pollutant concentrations at which an effect is observable.

The effects of *sulphur dioxide* have been studied for many years.[4] The gas acts by producing local injury, and if the duration of exposure is short the plant as a whole is not affected. Alfalfa, barley, cotton, and lettuce are all sensitive to levels of sulphur dioxide in excess of about 1.5 ppm for periods of an hour or more. By contrast, citrus fruits, cucumber, wisteria, and maple are relatively resistant to sulphur dioxide. The very heavy sulphur dioxide concentrations produced by industrial processes have led, as noted in chapter 2, to localized vegetation damage downwind from the plant. The

TABLE 6

AIR POLLUTANTS AND VEGETATION

POLLUTANT	SOME SENSITIVE SPECIES	APPROXIMATE CONCEN-TRATION AT WHICH EFFECT IS OBSERVABLE
Sulphur Dioxide	Alfalfa, Lettuce, Rhubarb, Spinach	0.5 ppm for four hours
Fluorides	Gladiolus, Apricot, Peach, Pine	1.0 parts per thousand million
Oxidants Particularly O_3	Tobacco, Apricot, Grapefruit, White Petunia, Spinach	0.02 ppm for ten hours
Chlorine	Similar to SO_2	0.25 ppm for one hour

This table gives an indication of some plants known to be especially sensitive to particular pollutants and the approximate concentration at which an effect has been observed.

careful studies of vegetation injury in the Sudbury area showed that, amongst the forest species, aspen, jackpine, and white birch were all relatively sensitive to injury from sulphur dioxide, being affected by concentrations as low as 0.2 ppm for eight hours; on the contrary, cedar, spruce, and maple were unaffected by these or higher concentrations.[68] These studies showed that the vegetation was more resistant to sulphur dioxide at night than during the daytime and more sensitive to effects in the months of June and July. As a result of these observations, it was suggested that, where emission of sulphur dioxide could not be avoided over such a rural area, periods of high emission should be curtailed during these months and during daytime, if that were possible.[68]

As noted in chapter 2, fluorides are emitted from industrial operations involving aluminum processing and the manufacture of phosphate fertilizers. The gas *hydrogen fluoride* is emitted most commonly, and is most injurious to vegetation. Many plants are exceedingly sensitive to this gas, particularly the gladiolus, some species of pine, and azaleas. Some varieties of peach, grapes, and strawberries are fairly sensitive, whereas resistant species include cotton, tobacco, alfalfa, beans, and citrus fruits. Episodes of specific damage have been reported from a number of different parts of the world, and litigation as a consequence of fluoride emission has occurred as a result of damage to citrus groves and to greenhouse plants and peaches. Fluorides may accumulate in plants which may be eaten by cattle, and in this way, and because of its very slow excretion, dairy cattle may gradually accumulate a dose of fluorine sufficient to cause changes in bones and joints, and evidence of fluorine intoxication.

The products of *photochemical smog* with their attendant oxidants have long been known to be harmful to vegetation. Ethylene, which occurs in smog in the Los Angeles region, is believed to have been responsible for injury to commercially grown orchids, and this gas in concentration below 0.1 ppm affects the growth of marigolds, sweet peas, and tomatoes.[4] The State of California has adopted an adverse level of 0.10 ppm for eight hours for this contaminant in its list of ambient air-quality standards. Other constituents of photochemical smog are amongst the major air pollutants with effects on vegetation. Oxides of nitrogen, ozone, and unstable peroxyacetyl nitrites together form a mixture to which many plants are sensitive. Many agricultural crops, fruit trees, and ornamental plants are affected by these substances, and the damage most commonly produced is white or brown flecks or blotches on the leaves.[4] These changes can be observed on grapes, citrus fruits, avocado, and on pine needles in the Los Angeles valley and surrounding hills, but the problem is not confined to this region of the world.

The studies by MacDowell[68] in Ontario showed that these compounds, probably predominantly *ozone,* had caused serious economic losses to tobacco growers in southern Ontario. As this author notes: "The flecking of tobacco leaves was clearly associated with hot, hazy weather, but air pollution was not suspected at first because the area was far away from industries and urban communities. However, observations and experiments reduced the possibility that the causal agent was a plant pathogen or a physiological factor, and when the air was analyzed the presence of photochemical air pollution was found. Ozone with very low concentrations was later identified as the phytotoxicant."[68] The highest ozone concentration that produced this damage was not more than 0.15 ppm, and a level of ozone of as little as 0.02 ppm for ten consecutive hours is believed to be the threshold above which injury may occur. The origin of the atmospheric ozone in the tobacco-growing region in Ontario was believed to be mainly waste products of petroleum combustion and refining in the United States.

Another effect of ozone is apparently to prevent citrus fruits from flowering properly, and there is little doubt that ozone in low concentrations greatly decreases the yield from such groves. Estimates of economic loss for the Los Angeles county region have been attempted, and have ranged as high as $5 million per annum for 1956. In economic terms, the secondary consequences of photochemical pollution appear to be greater than those attributable to other contaminants. Susceptible plants include petunias, some varieties of chrysanthemum, orchids, pansies, and zinnias. Amongst vegetables some varieties of beans, spinach, and lettuce are sensitive, whereas cabbage, tomatoes, and corn are all relatively resistant.

Hydrogen sulphide is usually not particularly toxic except in high concentrations, though some injury has been noted to trees and crops in the vicinity of certain natural gas wells in Alberta.[68]

Chlorine is about three times more phytotoxic than sulphur dioxide, but episodes of damage have occurred only when there has been an accidental leak of this gas as a result of railway accidents or explosions.

Many of the characteristic changes produced in plant leaves described in the foregoing section are well illustrated in the World Health Organization publication on air pollution.[4] Many of the lesions are quite characteristic, enabling immediate identification of the pollutant responsible. In most of the experimental work involving air pollutants and plants, a close interaction has been demonstrated between ambient temperature conditions, humidity, and stage of growth of the plant and its relative sensitivity to any given pollutant. This means that although the concentrations of a given pollutant may vary between rather narrow limits, periods of damage to vegetation may be evident only after high temperatures or high humidity conditions, or during certain months of the year.

EFFECTS ON ANIMALS

Apart from the ingestion by cattle and other grazing animals of vegetation containing high levels of fluorine, air pollution has very few effects on animals. It has been stated that the progressive blackening of tree trunks in industrial England has led to slow change, by natural selection, in the colouring of some species of moths which depend for their concealment on blending with the bark of such trees. It may also be true that some species of birds avoid heavily polluted industrial regions, but it is not easy to attribute such wisdom solely to their wish to avoid air pollution. It may be that long-term atmospheric effects could have some influence on life in the sea, and this problem is mentioned briefly in chapter 5. Although experimental animals have been used to test the effects of some air pollutants, there does not appear to be any information relating to the effects on the health of domestic animals of living in polluted cities.

The official report on mortality and morbidity during the London fog disaster of December 1952 noted in an appendix that there had been excess mortality in the cattle exhibited at the Annual Cattle Fair at Earls Court in London from 5 to 12 December.[61] Sheep and pigs were not affected, but the report noted: "The onset of the fog on the Friday (Dec. 5) was followed by acute respiratory symptoms in a number of the cattle, some 60 needing major veterinary treatment, and about 100 others requiring some form of minor attention. Twelve of the more serious cases were slaughtered, and one died. All were fat, in prime condition, and were relatively young. Autopsies on these cattle showed evidence of an acute bronchitis and some edema in the lung." The report notes than an anonymous article written in 1874 describes a similar episode affecting cattle at the Smithfield show during a fog which occurred between 9 and 12 December 1873. Though I have been unable to find a written recording of the observation, it was stated after the 1952 episode that the mortality had been higher amongst champion cattle, and this was attributed to the fact that the less favoured animals were in freer contact with their own excreta which, by giving off ammonia, protected them from the sulphur dioxide in the atmosphere.

EFFECTS OF AIR POLLUTION ON VISIBILITY

The reduction of visibility that occurs in the presence of high levels of atmospheric pollution must be ranked as one of the more important secondary effects of general urban air pollution. One of the most obvious consequences of a high level of particulate pollution is reduction in visibility and when this is aggravated or even caused by a high level of air pollution it must be considered an important contemporary hazard in aircraft operations. Each of the three main types of urban air pollution—namely particulate pollution, sulphur dioxide emission, and photochemical smog—has rather distinct effects on visibility, and for the sake of simplicity each of these will be considered separately.

Particulate pollution limits visibility in a more or less direct way in relation to mass concentration of the particles.[12] Some idea of this relationship may be given by the following numerical values. If the particulate concentration is approximately 50 $\mu g/m^3$, the visibility would be roughly ten miles. With a particulate concentration of 200 $\mu g/m^3$, the visibility would be cut to two miles, and in the range of 600 $\mu g/m^3$ the visibility would be less than a mile. It has been found that size distribution of atmospheric-suspended particles at ground level, when the number is related to particle radius, is reasonably constant. However, the figures just given would be affected by other aspects of the prevailing meteorological condition, particularly the relative humidity on the day in question. When this exceeds approximately 70 per cent, many particles grow into fog droplets. Naturally occurring particles, such as sodium chloride from an adjacent sea-coast, as well as man-made particulate pollution can act as condensation nuclei, and under these conditions the relationship between visibility and particulate pollution may be altered to a considerable extent. In general, the *higher* the level of humidity, the *lower* will be the level of visibility for any given level of particulate pollution; but there are some qualifications to be made to this statement that are too technical to be detailed in this volume.

Visibility in the atmosphere is reduced by two major effects that particles have on visible radiation. The first is the actual attenuation of light passing from object to observer, the result both of absorption and of scattering; and second is the illumination of the intervening air which results when sunlight is scattered into the line of sight by the particles lying between the observer and the object. This is the phenomenon that accounts for the fact that dark objects such as mountains appear to be progressively lighter in colour as they become more distant. In the modern urban environment in the presence of dry air and of pollution by very small particles, if an observer looks *into the sun* at a tall building one-quarter mile away he will observe that all detail on the building has been lost as a consequence of light scattering by the particles between him and the object (see plate 2). By comparison, if he turns and looks at a building with the sun behind him, he may find little evidence of particulate pollution to be visible.

The presence of sulphur dioxide in air gives rise to reduction in visibility by the formation of sulphuric acid mist and suspended sulphate particles. The relationship between gaseous sulphur dioxide and the formation of sulphuric acid mist is dependent on relative humidity, but under normal city conditions the sulphuric acid mist concentration may be considerable. A calculation has been made, for example, that in New York City, if the sulphur dioxide concentration is 0.3 ppm, the suspended particulate matter other than sulphuric acid will be approximately (0.3×1200) or 360 μg/m^3.[14] If the relative humidity were to be 90 per cent, this would result in a sulphuric acid mist content of 78 μg/m^3. Under these highly adverse conditions, the presence of these droplets *alone* would restrict visibility to 1.4 miles. At any given level of sulphuric acid mist concentration, the actual visibility is closely related to the relative humidity. Thus, if the sulphuric acid mist concentration were to be 100 $\mu g/m^3$, in the presence of a relative humidity of 50 per cent, the visibility would be roughly ten miles; but if the relative humidity were 98 per cent, the visibility would be approximately one mile. In terms of sulphur dioxide itself, at a concentration of 0.1 ppm, and with a relative humidity of 50 per cent, the visibility would be approximately seven miles. With the same concentration of sulphur dioxide, and a relative humidity of 98 per cent, it would be only two miles. Rises in humidity result in increases in the ratio of sulphuric acid mist to sulphur dioxide, accompanied by growth in the mass median diameter of sulphuric acid droplets.

The products of *photochemical smog* have separate effects. Nitrogen dioxide (NO_2) has a high coefficient of absorption of light in certain wavelengths. However, the concentrations of this gas are not high enough to be responsible for the considerable reductions in intensity which have been observed during intense photochemical air-pollution episodes. In one such incident, a reduction of as much as 80 per cent of light intensity at a particular wavelength has been observed at the earth's surface, but this was attributed to the light-scattering effect of atmospheric aerosols being formed as by-products of the photochemical reactions between hydrocarbons, oxides of nitrogen, and ozone. Although the presence of strong sunlight is known to be important for many of these reactions, it is likely that other factors, particularly the presence of carbon monoxide and a high ambient temperature, are also necessary for them to occur, and it seems probable that the ambient humidity also determines the effects on visibility of any given level of the products of photochemical smog. Very little is known about these factors at the present time.

Under actual conditions, the interaction between these various factors is very complex. Thus, in Manhattan, for example, the humidity may be high because of the proximity of the sea, the temperature may be high because of latitude of the city, sulphur dioxide is a common pollutant as a consequence of electrical power generation (and the demand for electric power in the summer is major in a city with air conditioning in common use), and the automobile density is high, producing the ingredients for photochemical smog. Under these extremely complex conditions, the precise effect of the combined air pollutants, particles, sulphur dioxide, and photochemical smog on visibility is not easy to predict. However, the combined effects of particulate pollution, oxidants, and sulphur dioxide on sunlight at the earth's surface have been carefully measured. Father Conrad East has shown[34] that these factors result in a 10 per cent reduction in incoming solar radiation in downtown Montreal compared with a station fifteen miles northwest of the city centre. Not unexpectedly, the attenuation is at a maximum during the winter months. Levels of attenuation of 7 per cent have been measured in Toronto, and values as high as 18 per cent were recorded in Boston between 1944 and 1948. It is hardly surprising that there are days in which the reduction of visibility probably attributable to air pollution interferes seriously with aircraft operations at La Guardia airport in New York.

By contrast, the problem in London, England, before the implementation of the Clean Air Act of 1956 was primarily one of large particulate pollution from open-burning of coal, high sulphur dioxide production, high ambient humidity, and a temperature inversion. Much of the sulphur dioxide under these conditions would exist as a sulphuric acid mist, and the particle size would be relatively large. The high humidity would lead to dense fog, the water vapour condensing around particulate aggregates and sodium chloride nuclei; the latter are particularly common in Britain since no city there is very far from the open sea. It has been a striking consequence of the passage of the Clean Air Act that the smoke concentration in London, which in 1958 was averaging more than 300 $\mu g/m^3$, by 1968 had been reduced to less than one-third of this level, with a corresponding considerable increase in hours of sunlight in the city (see chapter 6).

It is aesthetically important to keep air pollution to levels at which visibility is not obviously compromised. The psychological effects of living in a city with a high level of air pollution are impossible to quantify, but everyone can recognize that when one moves from such an environment into a rural area completely clear of air pollution there is an immediate sense of release and pleasure. In areas remote from modern cities, the particulate pollution may be as little as 10 $\mu g/m^3$. Many cities have average levels of twenty times this amount, and whether or not one attributes some psychological consequences directly to interference with visibility, such obstruction is surely one of the unaesthetic aspects of air pollution that deserve serious consideration.

EFFECTS OF POLLUTANTS ON MATERIALS

It has long been recognized that the constituents of air pollution have a deleterious effect on buildings and on other materials. There have been many attempts to quantify the cost of these expenditures, which in general fall into the following categories.

Effects on Metals

The rusting of steel occurs at a rate closely related to the dustfall of the region. In the United Kingdom, for example, iron specimens corrode four times faster in industrial areas than in rural areas, and in Britain the high levels of sulphur dioxide pollution have been held responsible for a major part of the corrosion of railway lines. The corrosion rates of metals are very much dependent on the concomitant presence of salt, and this factor must be taken into account when comparisons are made of the effect of air-pollution levels. Zinc is also affected by sulphur dioxide air pollution, and it has been shown that nickel and copper, although relatively corrosion resistant, corrode more quickly in industrial locations than in rural localities. In Britain, the Beaver Report of 1954 estimated the cost of metal corrosion as a consequence of air pollution to be $25 million per annum as a conservative estimate.

Effects on Building Materials and Painted Surfaces

The combination of sulphur dioxide and soot obviously has adverse effects on paint and on stone surfaces of buildings.

Houses in polluted areas require painting twice as frequently as those in clean areas, and the economic cost of more frequent renewal of paint in urban areas is considerable. For example, in France it has been estimated that these expenditures are well in excess of 15,000 million francs per year as a very conservative estimate.

In the United Kingdom the external maintenance of buildings in rural areas must be undertaken every three years, in moderately polluted areas about every two years, and in very polluted areas, every year. The cost incurred for cleaning and painting buildings has been estimated as between 15 and 40 per cent higher in polluted towns than in smokeless districts.

Effects on Fabrics and Dyes

The increased laundry costs associated with air pollution are considerable. The reduction in smoke accomplished in Pittsburgh was estimated to result in a saving of more than $6.5 million a year in laundry charges in that city alone. An important effect on fabrics and dyes is exerted by ozone.[13] Particularly when the fabric is wet, ozone attacks cotton fabrics by direct damage to the cellulose. Ozone also adversely affects the breaking strength of white nylon and polyester fabrics. Oxidants and oxides of nitrogen also cause fading of certain dyes, and a combination of high humidity and ozone has been considered responsible for fading of nylon carpets. Curtains are particularly vulnerable to air pollutants, and the increased wear necessitated by frequent cleaning as a result of particulate air pollution greatly limits their life. Curtains weakened by exposure to atmospheric soiling and acidity may be noted to split in parallel lines along the folds where the greatest number of particles have accumulated.

When sophisticated measurements of air pollution were not available, one could make a rough comparison between the particulate pollution of different areas by the days one could respectably wear a white shirt. This, however, was at a time when large soot-particle pollution was the commonest type, and the indication is of course not a reliable measure of such phenomena as oxidant air pollution; possibly the longevity of nylon stockings would now be a better indicator.

Other Effects

Oxidants have a major effect on rubber. Ozone causes cracking of rubber unless it is protected by compounds reducing the oxidant effect. The behaviour of rubber exposed to ozone under experimental conditions is broadly related to the service behaviour of tires in localities where atmospheric ozone is high, and the rate of cracking of the rubber is primarily a function of the ozone concentration. Antiozonant additives have been developed to protect rubber against ambient ozone, but these chemicals are expensive and add materially to the cost of the product.

ESTIMATES OF ECONOMIC COST OF EFFECTIVE AIR POLLUTION ON MATERIALS

Many estimates have been made of the approximate cost of air pollution to the community with respect to all of these factors. A study made in France in 1957 put these expenditures, excluding any health charges, at approximately 4500 francs per inhabitant per year. Estimates in Britain for 1947 put the cost at approximately £100 million per annum, or £2 per inhabitant per year. It was calculated that the emission of one ton of smoke caused damage to the approximate value of £20.

A careful study of differences in cleaning costs incurred at two U.S. towns compared mean annual suspended particulate pollution of approximately 60 μg/m³ in one (virtually rural) with approximately 280 μg/m³ in the other (heavily polluted). The results suggested that the outlay for maintenance of materials in the latter town was higher than the former by a factor of roughly four. Comparison between two other towns, one with a mean annual suspended particulate pollution of about 110 μg/m³ and one running at about three times that level, led to the conclusion that the differences incurred between the two in reference to the outside and inside maintenance of houses, in laundry and dry cleaning, and in personal hygiene costs amounted to $84 per capita per annum.

Inaccurate though these estimates may be, they are based on careful comparisons between heavily polluted and non-polluted regions, and must be considered to give broadly correct general indications of costs incurred from air pollution. The relevance of these observations to decisions concerning air-pollution legislation is discussed in chapter 6.

chapter four

EFFECTS OF POLLUTANTS ON MAN

INTRODUCTION

The suspicion that air pollution might be having an adverse effect on health is a very old one and extends back two or three hundred years. Before the days of antibiotics and in times when medical treatment was primitive and life expectation was short, no meaningful attempt could have been made to measure the contribution, if any, that air pollution might be making to mortality. Many people, however, are surprised to learn that it is still very difficult to give precise answers to the obvious questions one can ask concerning the influence of different kinds of air pollutants on human health. The reasons for this difficulty are multiple, but it is essential that the reader should have some idea of the problems which exist in this area of study so that he may be on his guard against oversimplified statements. The latter may do something to illuminate the issues, but they do little to illustrate the truth. Some of the reasons why this area of study has proved particularly difficult are the following.

1.   In many countries the calculation of pollutants has been confined to crude surveys of dustfall, or to measurements of total particulate pollution and sulphur dioxide. The reader will realize that these two measurements may summarize, in a general way, the level of pollution of a region, but they do not of themselves constitute a very complete itemization of potentially harmful substances which are not infrequently found in the air. Furthermore, if some relationship is demonstrated between an increased incidence of a particular illness and higher levels of these two pollutants, there is a tendency to assume that the pollutants themselves have been responsible for the higher incidence, and this may be incorrect.

2.   Since 1920, there has been a very considerable rise in the number of cigarettes smoked by the population. The impact of this increase on lung cancer was not documented with any degree of certainty until 1952. It had become clear by 1956 that cigarettes were playing a major part in the production of chronic respiratory diseases, particularly chronic bronchitis and emphysema, in addition to their role in causing lung cancer. Since chronic bronchitis and emphysema (which are described in detail later) are also possibly related to air pollution, it became essential to be able to separate the effects of cigarettes from those of air pollution; in some early studies this distinction was not made.

3.   There are considerable individual variations in sensitivity to substances like sulphur dioxide and ozone, and therefore in a civilian population one would expect wide differences in sensitivity amongst different people. Unless concentrations of substances reach such a level that a major fraction of the population is overtly affected, it may be very difficult to protect the few individuals who have in fact been adversely affected.

4.   Although a number of studies have indicated that certain diseases occur more frequently or appear to be more severe in cities with high air pollution, there are other factors that may be related to this. For example, there might be a potentiation of the growth in the presence of air pollutants of some bacteria or viruses that attack the human respiratory system. Or there might be important effects due to high popu-

lation density which are not directly related to air pollutants, but have to do with the greater likelihood of encountering people with a respiratory illness in public transportation and offices, whereas the dweller in the country is in much less frequent contact with other people. Such factors are difficult to randomize in comparative studies.

5.   Finally, there are important variations in climate and temperature which of themselves may affect human health and produce increases in mortality or in illness. In regions such as Los Angeles, episodes of high temperature may be associated with the production of more photochemical smog, but in addition the high temperature itself is likely to lead to increased strain on those who already have existing disease of the heart or lungs. In colder climates, episodes of fog and low temperature may be harmful to people with pre-existing lung disease, quite apart from the fact that these climatic conditions may commonly be associated with high levels of air pollution. It is important therefore to attempt to separate the effects of climatic changes from those of pollutants in any studies designed to illustrate the relationship between disease and air pollution.

As if these difficulties weren't severe enough, there are others to which attention has recently been directed. There may be in certain populations a predisposition to particular types of ailments, and it could be that some of these might be misinterpreted in terms of air-pollution effects. There are formidable practical difficulties in such studies, amongst which one might include the difficulty in a modern city of obtaining quick and up-to-date information on the admission of patients to large numbers of hospitals over specific time periods. There have been few attempts so far to link together all the hospitals of a major metropolitan area, so that on an hour-to-hour basis admissions of people in acute respiratory difficulty might be tabulated. If a major city contains 30 hospitals to which such patients might ordinarily be admitted, it is possible for each of those hospitals to see three or four such patients between the same two-hour period in the afternoon, and although this would not appear particularly dramatic in the individual hospital setting it might indicate a considerable occurrence of general effects of an air-pollution episode. Furthermore, although such acute episodes may be part of the pattern of effects of air pollution, the very long-term consequences are for different reasons equally difficult to study. Suppose an air pollutant were to have a very slow effect on the lungs, visible only after twenty years of residence in that environment, what chance would there be of detecting it?

This by no means exhaustive account of the difficulties which confront the physician or epidemiologist trying to obtain some firm answers to what are, after all, very important but very simple questions may go some way to explaining why hard and fast data have been difficult to come by. Indeed, it is the observation that diverse kinds of evidence point generally in the same direction that leads to the conclusion that there have almost certainly been considerable effects on human health by existing levels of pollutants. There is no single study which illustrates this conclusion with as much force as does a consideration of all the kinds of evidence which point in this direction.

It is sometimes said that no disease has been shown to be directly attributable to air pollution, and in a sense this is true. It is however beyond dispute that air pollution has contributed to human ill health, though the magnitude of this effect is very difficult to quantify.

The literature on health effects is considerable, and many studies contributed from a number of different countries in the past five years have done much to improve our knowledge of this area. Only a small fraction of this literature can be summarized, and in the succeeding pages I have dealt at greater length with those studies which appear to me to be most carefully controlled, and least likely to have been influenced by factors other than the variable of air pollution. The references contained in the text will enable the reader to delve further into this complex subject if he wishes.

THE LUNG AS A TARGET ORGAN

For the reader to be able to understand the implications of studies of possible interrelationships between illness and levels of air pollution, it is essential that he should have some knowledge of the defences of the normal lung and of the different kinds of illness which have been the subject of investigation. In this book, a full description of these disease entities and physiological functions is not possible, but the following brief accounts should help to guide the reader in the later discussion.

The Natural Defences of the Lung

It is easy to forget that the lungs, which in a normal man have an internal surface area of 70 square metres, are continuously exposed to the external environment. The branching airways, which start as a single tube about 2.5 cm in

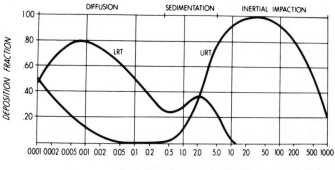

AERODYNAMIC DIAMETER (MICRONS)

FIGURE 4.   DEPOSITION RATE OF PARTICLES IN THE HUMAN LUNG

This figure shows the fraction of inhaled particles which will be deposited in either the upper respiratory tract (URT) or the lower respiratory tract (LRT) in human subjects. Notice that almost all particles between 10 and 100 microns in size will settle in the upper respiratory tract; many of these will be retained in the nose. It is also apparent that an increasing fraction of particles between 0.5 and 0.01 microns will be laid down deep in the lung in the lower respiratory tract. Particles smaller than 0.01 of a micron have a lower deposition rate in the lung. The aerodynamic diameter of the particle refers to its actual diameter which, in the presence of water vapour, may be rather larger than the physical size of the particle in dry air. (Figure courtesy of Prof. Paul Morrow, University of Rochester, N.Y.)

diameter, end as ducts only half a millimetre in diameter, but there are more than half-a-million of these. The air sacs, called alveoli, are very thin in order to permit fast diffusion of oxygen from gas to blood; they lie at the end of the airway system.

The entire respiratory system possesses formidable defences. The *nose* is very efficient at trapping large particles, and prevents particles between 20 and 50 microns in size from getting into the lungs (see figure 4). Not only this, but soluble gases such as sulphur dioxide are almost completely removed from air as they pass through the nose. The nose and the *large airways* possess on their inside surfaces small cells with motile "hairs" called cilia. These, by synchronized movements similar to a wind passing over a field of wheat, propel particles upwards on a blanket of sticky mucus. This important clearing mechanism is capable of removing considerable quantities of dirt from the lungs, but its efficiency and activity are impaired by such factors as excessive dryness or irritant smoke. The airways possess many sensors that can stimulate a cough. This forceful propulsive mechanism can drive collections of mucus or large inhaled materials, such as pieces of nut, out of the airway into the mouth.

Deep in the lung, wandering cells called *macrophages* can pick up and digest or, if that is impossible, store particles and remove them from the alveolar surface. Although there are many aspects of the operation of these cells that remain to be elucidated, it seems clear that they represent a very important defence line against bacterial invasion. They probably remove particles of carbon or silica through the walls of the alveolus, and such particles may finally end up being stored, for the life of the individual, in *lymphatic glands* situated at the root of the lung (see plate 3). It is clear that particles more soluble than these may end up in the blood stream, in which situation later clearance may occur through the kidneys. This is presumably the route whereby small inhaled particles of lead from automobile gasoline pass through the lung, and are finally removed from the body.

The macrophages contain and, in the course of their work, may release enzymes capable of destroying protein. The lung has defences against the effect of these enzymes on lung tissue itself, and it seems clear that this system is important for the protection of the lung against internal damage.

Figure 5 shows the volume of small particles one would predict would be deposited within the lung under specific conditions of ambient pollution, particle size, and physical activity.

Some Relevant Diseases

Chronic bronchitis is a chronic change in the airways of the human lung often complicated by bacterial infection, but believed to be most commonly caused by irritation from cigarette

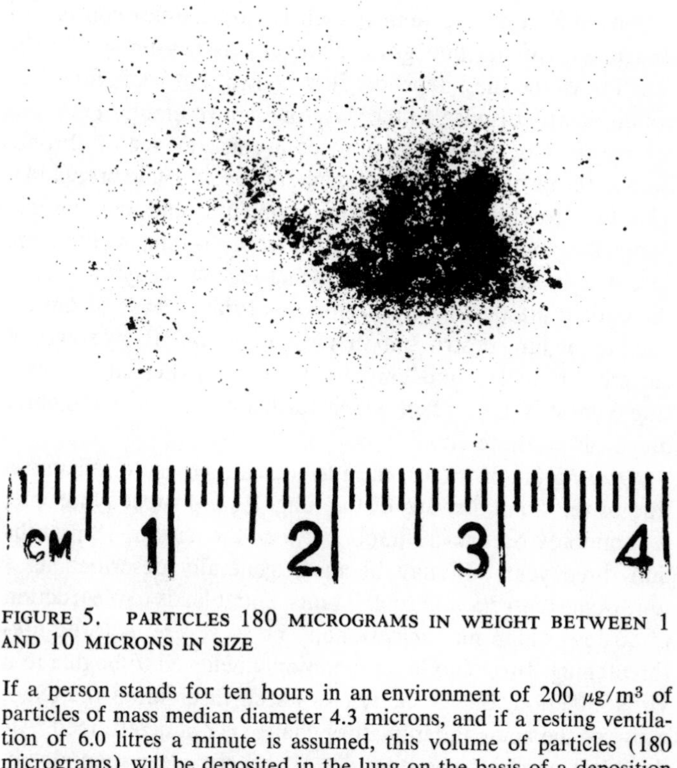

FIGURE 5. PARTICLES 180 MICROGRAMS IN WEIGHT BETWEEN 1 AND 10 MICRONS IN SIZE

If a person stands for ten hours in an environment of 200 $\mu g/m^3$ of particles of mass median diameter 4.3 microns, and if a resting ventilation of 6.0 litres a minute is assumed, this volume of particles (180 micrograms) will be deposited in the lung on the basis of a deposition rate of 25 per cent computed from figure 4. If the ventilation were to be doubled, as it would be for instance by light exercise, this deposition rate would be doubled.

smoke or environmental influences. It brings on a chronic cough often productive of phlegm and, if the changes in the airways have led to an increase in resistance to airflow through them, there may be some shortness of breath associated with the condition. The disease is said to be present when there is a history of a chronic cough (usually best evaluated by means of a questionnaire), and when no other illness is responsible for this symptom. Chronic bronchitis is known to be a common consequence of long-time cigarette smoking; from being a relatively simple benign condition, it may progress to a severe form, or may be followed by the advent of destruction of the fine air sacs in the lung, leading to emphysema.

It is known that there are multiple causes of *emphysema*, and some different kinds of the condition, but in essence this serious change in the lung (which is irreversible) consists of destruction of the fine gas-exchanging tissues on which the function of the lung depends. It is known that some inherited conditions predispose to the development of emphysema, but the majority of cases appear to be a consequence of chronic bronchitis, or of the same factors that cause chronic bronchitis. The two conditions of chronic bronchitis and emphysema merge into one another, and it is far from easy during life to be sure whether they coexist, or to what extent the symptoms of the patient are dependent on disease of the airways or are related to the lung destruction that is emphysema. Early stages of chronic bronchitis and emphysema are not detectable on a single chest X-ray, a fact which further complicates the early diagnosis of these conditions.

Bronchiolitis is a term used to indicate an acute inflammation of the small airways of the lung. As a distinct syndrome, it commonly occurs in infants between the ages of 18 months and three years. It may be more generally described as a *lower-chest infection* in such infants, and it leads to obstruction of airflow which may occasionally be so severe as to be life-threatening. Bronchiolitis is commonly believed to be due to a virus infection, but in many cases seen in children the exact cause of the acute inflammatory change is not determined. Although it carries some mortality, it usually clears completely leaving as far as is known no harmful aftereffects.

Asthma is a disease in which obstruction to the airways is spasmodic, and believed to be caused usually by sensitivity to some protein with which the person has come in contact. Sometimes any wheezing in the chest is designated as asthma, but accentuated spasm may occur in association with, or as a result of, chronic bronchitis.

Lung cancer in adults has been shown to be very significantly related to cigarette smoking. In the past twenty years, it has become a disease of major health concern, and is now the commonest form of cancer affecting men in the western world. There are different types of lung cancer, not all of which are believed to be related to cigarette smoking, but the insidious nature of the condition and its tendency to spread make it particularly difficult to detect and treat in the early stages.

Pulmonary fibrosis is a term that includes many kinds of change in the lung, and different inhaled materials such as coal dust, silica, beryllium, and asbestos all produce the changes which may be described as "fibrosis." These conditions, however, are very different when examined by the pathologist. There also occur patterns of pulmonary fibrosis that, as far as we know, are not related to any material with which the subject may have been in contact. All of the materials mentioned above are known to be capable of causing occupational lung disease, and exposure to them may lead to pulmonary fibrosis.

It is important for the reader to understand that the individual sensitivity of people to substances they encounter in the environment varies greatly. This is obviously true of cigarettes, since one meets individuals who have smoked a large number of cigarettes over a long period of time and who have little in the way of significant chronic bronchitis, no pulmonary emphysema, and no lung cancer. Others, however, may be affected adversely by exposure to cigarettes over a much shorter period.

The same is true of such substances as asbestos and silica, and one must presume, in the absence of other evidence, that there is a wide individual variation to be expected in the response of an average population to such items as general particulate pollution, sulphur dioxide, oxidants, and any other material to which people may be exposed. This point is considered in some detail later in chapter 6.

MORTALITY DATA

The first attempts to examine the effect of urban living on health were essentially studies to see if deaths from different diseases varied in different geographic regions. Such investigations suffer from two limitations: firstly, the precision with which the cause of death may be certified may not be very great, and may vary between different regions of the same country as well as between different countries. Secondly, mortality, whether acute or chronic, represents an end point (!) and is therefore inevitably a crude indicator of the total burden of ill health which may be associated with modern urban living. Nevertheless, there have been significant associations between air pollution and mortality and these deserve consideration first.

Acute Episodes

The most dramatic demonstrations of the effects of high levels of air pollution on health have been the acute episodes that have occurred over the years. An instance of acute industrial air pollution in a valley in Donora, Pennsylvania, during October 1948 resulted in an attributed mortality of approximately 20.[4,5] An earlier episode in 1930 in the Meuse Valley, Belgium, had a somewhat higher mortality and was considered attributable to very high concentrations of some specific pollutants. In 1950 in Poza Rica, Mexico, an acute episode of air pollution was thought to be caused by the accidental emission of hydrogen sulfide gas.[5]

Such events are not really comparable in any way to the common levels of pollution encountered in modern cities. But it was the London disaster in December 1952 which showed beyond any doubt that a very large excess mortality might be produced in a modern city if a prolonged temperature inversion occurred with considerable emission of sulphur dioxide and smoke. Figure 6 shows the levels of smoke and SO_2 during this episode, together with the increased number of deaths occurring in the greater London region. This incident has been fully described in the literature, [7,61] but it was on an entirely different scale from previous phenomena. The victims were mostly over the age of 45 and there was little doubt that many of them suffered from pre-existing lung disease or heart disease, but there was also noted to be an increase in infant mortality, though the number involved was small. The greatest single increase occurred in deaths certified as being due to

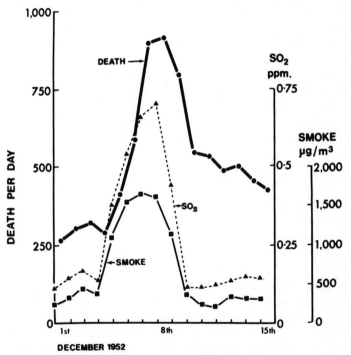

FIGURE 6.   AIR-POLLUTANT LEVELS AND EXCESS MORTALITY IN LONDON IN DECEMBER 1952

This diagram has been redrawn from reference 7. Notice that over a period of three days, sulphur dioxide exceeded 0.5 ppm and the smoke level exceeded 1000 $\mu g/m^3$. These figures should be compared with data given in figure 1 and table 5. During this episode, 4000 excess deaths occurred over and above those normally expected.

"bronchitis." These increased approximately tenfold between the week before the fog episode and the week of the fog. There appeared to be few deaths amongst healthy adults that could be attributed directly to the episode, and none of these were very precisely documented.

After this disaster, much more attention was given to seasonal changes in mortality in London and it was possible to show that increases in death rate occurred with each abrupt change in pollution in succeeding years, particularly when the smoke density increased above 750 $\mu g/m^3$ and with the sulphur dioxide above 0.5 ppm. During a 1962 episode in London, on almost exactly the tenth anniversary of the first one, excess deaths were believed to have numbered about seven hundred, with again the greatest proportionate increase being in bronchitis.[7]

Significant increases in mortality have been shown to occur in other major cities when there is a severe temperature inversion. This was found to be true of the New York City area during the period 12–21 November 1953, and subsequent studies have indicated just detectable effects in other episodes.

Long-Term Mortality Data

In respect to deaths from bronchitis and some other conditions, there is clearly some general relationship between air pollution and this illness as a cause of death. Considerable difficulty exists in standardizing the certification of causes of death, but even when these uncertainties are set aside it is beyond argument that very high levels of air pollution, judged from the point of view of particulate and sulphur dioxide pollution, are, in a general sense, associated with higher mortality rates for bronchitis. This has been shown to be true of different parts of the greater London area, of different regions of Buffalo (New York) and its suburbs, and of different industrial and rural areas in Britain.

It has become clear that this disease in particular is commoner in the severe form in the cities of Britain than elsewhere in Europe, and a good deal commoner in Britain than in North America;[53] it is far from certain why this should be.

If the mortality from bronchitis is plotted on a graph against the crude pollution measured only in terms of particulate pollution, provided enough regions are studied, a recognizable general relationship emerges. All these data taken together indicate that, at least in the climate of Britain and probably to some extent everywhere, living in a region with a high SO_2 and smoke level is associated with a substantial increase of mortality from chronic bronchitis.

The interpretation of data relating to lung cancer and air pollution is exceedingly complex. There are many difficulties in being sure of the precision of diagnosis, the satisfactory consideration of the cigarette-smoking history of the individual, and in the separation of air-pollution effects from those that may be related to the occupation of people who live in a particular area. Major studies are now in process to obtain better information on this point, and we will have to await the completion of these before a definitive opinion can be given; however, enough data now exist for it to be clear that in relation to the preponderant effect of cigarettes the contribution of air pollution to lung cancer levels is small. The Los Angeles population, which is exposed to a kind of air pollution different from that existing in Great Britain, does not apparently have any excess mortality from lung cancer by comparison with other cities in the United States. There has been some evidence that cancer of the stomach may be related to levels of air pollution, but these data are essentially preliminary and must be considered unproven at this point.

It might be supposed that a careful comparison of data secured from autopsies might give an indication of comparative incidence of chronic bronchitis or emphysema in different populations. Chronic bronchitis is not easy to quantify from an autopsy of the lung, but some attempt can be made to examine by a standardized method the occurrence of the destructive lesion of emphysema in autopsy material. One group of pathologists has compared the occurrence at autopsy of emphysema in Winnipeg in Canada with St. Louis in the United States. Although precise correlation of cigarette smoking was not possible, the findings indicated that the incidence and severity of emphysema were greater in St. Louis (which has a very much higher level of air pollution) than in Winnipeg, which is almost pollution free. These data provide some evidence of a difference in frequency but it is hard to be sure that all other variables have been excluded in an autopsy study of this kind.

From many points of view the attempt to study the effect of air pollution using only mortality data is unsatisfactory, and for this reason many more studies have been attempted in the past few years of the effects on health as a whole. These are discussed in the next section.

STUDIES OF MORBIDITY

Chronic Respiratory Disease in Adults

Studies of the effect of air pollution on chronic bronchitis in the adult population are made difficult by virtue of differences in cigarette-smoking habit, differences between occupations of men who live in rural and urban areas, and the possible impact of differences in economic status which, by leading to different levels of nutrition or of medical care, might bring about a higher incidence of chronic lung disease in lower income groups than in the more privileged members of the community. A number of early studies attempting to compare populations did not take sufficient account of these factors. It has now become clear that to attempt any soundly based answer to the question of whether existing levels of air pollution have influenced the occurrence of chronic disease of this kind, it is necessary that very full and careful account be taken of influences in the population as a whole which may affect the frequency of occurrence of this common disease. Furthermore, as has been pointed out above, the early diagnosis of these conditions is not a simple matter.

When these qualifications have been made, however, there exists a considerable body of information suggesting that a factor is present in cities with relatively high levels of sulphur dioxide and particulate pollution which leads to a greater incidence of chronic bronchitis and emphysema. In Great

Britain, a higher occurrence of chronic bronchitis, particularly in men, in major urban areas such as London has been clearly demonstrated.[7] It has been shown that there are generally higher rates for bronchitis in urban than in rural areas when cigarette smoking has been fully taken into account; if rural areas are compared with major cities, in urban non-smokers there is a significantly higher occurrence of bronchitis. However, it is in moderately heavy cigarette smokers that the greatest increases in bronchitis appear to occur if the man lives and works in the urban environment.

A careful comparison of men employed by the British post office to drive postal vans in and around three country towns in England (Gloucester, Norwich, and Peterborough) with an economically similar group in London, taking full account of differences in cigarette smoking, showed that the London group had more respiratory symptoms, a greater production of phlegm, and a measurably greater impairment of ventilatory function than did the workers in the country towns.[52] The authors of this study concluded: "Of the factors reviewed in this study, differences in local levels of air pollution appear to be the likeliest cause of the differences in respiratory morbidity between men working in central London and those in the three rural areas."

In the United States, the level of chronic bronchitis in major urban areas appears to be generally lower than in Britain, though there is a similar increase in the occurrence of this disease as between men who work in a major urban centre and men who work in a small town or rural environment.[53] A number of correlations between Great Britain and the United States have been undertaken, and these have tended to show that the occurrence of severe chronic bronchitis, amongst men of comparable economic circumstances and smoking habit,[53] is commoner in London than in cities on the eastern seaboard of the United States. On both sides of the Atlantic, an additive urban factor to cigarette smoking is demonstrable.

In Holland a random sample of men and women aged between forty and sixty-five in an industrial town, with a considerable air-pollution problem, and in a semi-rural area showed that for all smoking categories persistent phlegm was significantly higher in the industrial city.[77] These differences could not be attributed to social or economic circumstance. In the United States, detailed studies of respiratory disease morbidity were undertaken over a period of years in the area of Buffalo, New York. Although in earlier studies detailed account was not taken of the smoking population, later observations have considered this factor, together with occupation. A general relationship between the occurrence of chronic bronchitis and emphysema appears to be demonstrable in this region in relation to the particulate and sulphur dioxide pollution which exists in the region as a whole, but the magnitude of the effect is smaller than in Great Britain.

In Genoa, Italy, advantage was taken of the fact that the city is divided by a series of ridges running down to the Mediterranean, giving rather different levels of sulphur dioxide and particulate pollution in different regions of the town.[66] After twelve air-sampling stations had been operating for three years, a study was undertaken to see if the levels of sulphur dioxide and particles bore any relationship to the occurrence of acute respiratory infections in different population groups. The results showed a much higher incidence of bronchitis in people living in regions where the average SO_2 was much above 0.04 ppm and the particulate pollution level above 180 micrograms per cubic metre. Infections involving the upper airways did not seem to be related to the pollution level in the groups studied.

In Japan, the levels of industrial air pollution are among the highest in the world, and a much higher occurrence of chronic cough, of chronic bronchitis, and probably of emphysema has been documented in regions with greater air pollution. The high levels of pollution in the Yokohama region were believed to be responsible for a type of chronic bronchitis characterized by attacks of acute chest wheezing that led to the condition originally being designated as "Yokohama asthma." The ailment was particularly evident, however, not in those with an allergic background or family history, but in those who smoked cigarettes as well as living in that particular region.

Although some attempts have been made to make comparisons between statistical data within Canada in terms of the incidence of chronic bronchitis and emphysema in different parts of the country, the diagnosis of these conditions in a routine manner is imprecise; it has not proved possible to determine whether or not there is yet any basis for comparing the incidence of these diseases between one Canadian city and another. No detailed epidemiologic studies have yet been attempted of rural versus urban populations in Canada in terms of chronic bronchitis.

It would not be surprising if the occurrence of an accentuation of air pollution produced a worsening of symptoms in people who already had chronic bronchitis and emphysema and who were living in that city. In the years following 1952, it was shown quite clearly in London that episodes of increased air pollution, particularly when associated with fog, were accompanied by a worsening of symptoms amongst a group of patients who already had chronic bronchitis and emphysema. It is interesting and encouraging that since the improvement in the London atmosphere as a result of the passage of the Clean Air Act of 1956, this relationship has become progressively more difficult to demonstrate[7] (see chapter 6). Similar accentuations of admissions to hospital for chronic respiratory disease have been noted in New York City during the occurrence of periods of temperature inversion, and such an effect is likely to be demonstrable in any city with a large enough population whenever the air pollution becomes more severe than usual.

Another approach has been to study the absence from work in relation to pollution. It has been shown in London that absenteeism because of bronchitis is measurably higher amongst those working in the most polluted areas when other factors have been taken into account. It was even found that amongst postmen and bus crews in London, there were higher rates of disability from bronchitis among men employed in the centre and northeast of London where the air-pollution levels are highest, as compared with southwestern boroughs in which levels are somewhat lower because of the prevailing southwest wind. Sickness rates from bronchitis in men aged 55 to 59 showed an inception rate for bronchitis of 94 per 1000 men in regions of high winter smoke pollution, compared with 59 per 1000 men in regions where the air pollution was designated as "moderate."[7] Surveys completed by the Royal College of General Practitioners in Britain demonstrated that after all other factors have been taken into account, chronic bronchitis and its acute exacerbations appeared to be approximately twice as frequent in the large towns of Britain as in the rural areas.

This information may be summarized by saying that levels of sulphur dioxide of more than 0.04 ppm as an annual mean, or more than 0.1 ppm as a daily maximum, seem to be associated with higher morbidity rates for chronic bronchitis in adults, and this effect appears to be most prominent in those who are cigarette smokers. Acute accentuations of pollution with increases of sulphur dioxide much above 0.25 ppm or particulate levels above 750 micrograms per cubic metre during an episode, provided this lasts for one or two days, are accompanied by a measurable increased morbidity and sometimes an increased mortality from respiratory disease, particularly amongst older people or those who already suffer from chronic bronchitis and emphysema.

There is therefore a very general consensus of evidence that an urban factor exists which increases the sickness rate from bronchitis in the general population. This effect is particularly exerted on those who are cigarette smokers though it is usually measurable to some extent in those who do not smoke cigarettes. This result is not unexpected since the inhalation of cigarette smoke which contains particles and oxides of nitrogen, amongst many other things, may be regarded as not essentially different from the more general inhalation of particles and sulphur dioxide and oxides of nitrogen (though these are in very much lower concentration) in modern urban air. Thus it is perhaps not surprising to find, in almost every study which has been done, that the differences between the urban and rural populations appear to be greatest amongst those who are moderate or moderately heavy cigarette smokers. Yet it is important to recognize that in non-smoking adults who have been studied the same accentuation of respiratory symptoms and some slight decrement of pulmonary function also seem to exist in the urban as compared with the rural population.

It may therefore be stated with confidence that residence in a modern urban atmosphere does potentiate the damage being caused by cigarette smoking and leads to more absence from work from bronchitis. In the damp climate of Britain, this seems to play a considerable part in determining the higher percentage of men with more severe chronic bronchitis seen in that environment as compared with other places. Possibly the interaction between climate and air pollution is important, but there are insufficient studies to enable one to assign some percentage of responsibility to climate versus the level of air pollutants.

It is far from clear, however, whether there is any evidence that living in the modern urban environment in fact leads to the development of chronic bronchitis in a non-cigarette smoker not exposed to any irritants in his occupational environment; indeed it is often not recognized that this is a different question. Although it may be admitted that aggravation of *pre-existing* change in the lung is attributable to living in an urban environment, and that the most likely factor is air pollution, the occurrence of bronchitis in the non-cigarette-smoking groups living in the urban environment is not very high. All the evidence indicates that cigarette smoking has a far greater effect on these diseases than does any urban factor, including air pollution.

However, the occurrence of chronic bronchitis is a matter of major concern whether viewed from the point of certified mortality or as measured by approximate indices of morbidity. In Canada, between 1950 and 1966 for example, the death rate per 100,000 of the population from emphysema and chronic bronchitis in men increased from 2.0 to 13.9, and the morbidity of chronic respiratory disease as judged by hospital certification increased from approximately half-a-million hospital days to nearly one-million hospital days over the same period. What fraction of this considerable and increasing burden of chronic respiratory disease can be attributed to cigarettes and what fraction to air pollution? It is important to recognize that no precise answer to this can be attempted at the present time. The question certainly deserves an answer since more and more of the population are moving into cities and control measures take some years to have any effect. The report of the Royal College of Physicians in London[7] concluded: "There is some evidence that air pollution is more important in the aggravation than in the initiation of chronic bronchitis."

Respiratory Disease in Children

It may be argued that if the only effect of air pollution on human health is to cause a slight worsening of diseases primarily attributable to cigarette smoking, then there is little reason to urge control of air pollution on grounds of human health. Because many have held such a viewpoint, the recently published data concerning the differences in respiratory-disease morbidity or pulmonary function in school children in relation to air pollution assume major significance. It may be suspected, though it is not provable, that exposure to air pollution in early years of life might have an adverse effect on the development of the lung, or by some mechanism may prepare the ground for the later development of chronic bronchitis if other circumstances favour this. It cannot be assumed that there is no relationship between levels of illness noted in school children and later cases of chronic bronchitis observed in adults.

There have now been several major studies of the relationship of respiratory symptoms or episodes of lower-chest infection in children in relation to the area in which they live. It has been shown that the prevalence of respiratory symptoms as well as a decrement in pulmonary function is present in children in Sheffield in relation to others living in areas of lower air pollution. In a study of school children between the ages of six and eleven, the increased morbidity of chronic cough was demonstrable only in children from families of less skilled workers (social classes 4 and 5).[7] In these children, the standardized morbidity ratio referring to the frequency of chronic cough was 95 in children from rural districts, 126 in children from Bristol and Reading (considered to have moderate or low levels of air pollution), and 149 in two cities with high levels of air pollution (Newcastle and Bolton).

In another study, data were collected on more than 10,000 children attending school in four areas of Kent, two being predominantly urban and two predominantly rural.[51] A significant relationship was demonstrated between the performance of a pulmonary-function test and symptoms of cough and phlegm, and both of these appeared to be affected to some degree by the area of residence of the child, the sector showing the lowest level of peak expiratory flow and the highest rate for respiratory symptoms having a higher level of air pollution than the other three. From Japan have come reports of a higher prevalence of respiratory symptoms and a lower performance of a simple expiratory test of pulmonary function in school children living in heavily polluted regions.

In 1966, a British study was published which dealt with a younger age group.[32] From a large sample of children being adopted at birth into other families, a total of 3,866 children was followed for a period of five years. By studying only children being adopted into other families, it was hoped that any familial genetic factors could be considered to have been randomized. The children lived in widely scattered parts of the country and the areas in which they were being brought up were classified into four general groups in terms of the air pollution existing therein. Information on illnesses was obtained from health visitors who interviewed the mothers of the children, and a careful note was made of the occurrence of upper- and lower-inspiratory infections. In table 7 are shown the general levels of air pollution in terms of smoke or particulate pollution and of SO_2 in the four designated regions of this study. Figure 7 illustrates the reported incidence of a lower-chest infection or bronchitis occurring in the first two years of life in relation to air pollution. It is clear that a nearly threefold increase in such illness was noted among those in moderately polluted zones as opposed to those in zones of very low pollution.*

*The reader may note by comparison of table 7 with figure 1 and table 5 that Montreal or Toronto would be classified by the criteria of table 7 as a zone of moderate air pollution. It is not known whether increased morbidity from lower-chest infection is occurring in these cities by comparison, for example, with Kingston.

TABLE 7

AIR-POLLUTION RATING

Smoke		Very Low	Low	Moderate	High
London (Mean)	$\mu g/m^3$		91	99	142
No. of areas			(8)	(6)	(27)
Other (Mean)	$\mu g/m^3$	67	138	217	281
No. of areas		(17)	(56)	(51)	(14)
Sulphur Dioxide					
London (Mean)	ppm		.055	.067	.092
No. of areas			(8)	(6)	(24)
Other (Mean)	ppm	.033	.048	.071	.095
No. of areas		(14)	(46)	(43)	(10)

These data are taken from the study (reference 32) dealing with the occurrence of lower-chest infections in infants. The number of stations in each category is indicated in parentheses under the observed levels of sulphur dioxide and smoke. It will be noted that the zones in Great Britain classified in this study as "moderate" have average levels of particulate pollution and sulphur dioxide comparable to those observed in city areas in Canada (refer to figure 1 and table 5). This table should be studied in conjunction with figure 7.

The same study also revealed that upper-respiratory infections such as tonsillitis or ordinary colds were not related to the air-pollution grouping, and the economic circumstances of the family into which the infant had been adopted had similarly no effect on the occurrence of chest infections. The care and thoroughness with which this study was completed means that its conclusions cannot be lightly dismissed. It suggests strongly (and until conflicting evidence is produced, must be taken to have established the fact) that the level of air pollution as indicated by these two indices is associated with the occurrence of an increased incidence of lower-chest infections in the first two years of life. Such a conclusion is in many ways surprising, since one would have thought that for much of the time such infants would be indoors where the levels of pollution might be expected to be less than in the environment as a whole.

When comparisons are made between the air-pollution levels in which these infants were being brought up and the general levels of pollution existing in North American cities, it

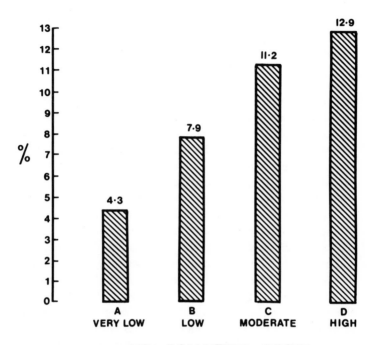

FIGURE 7. OCCURRENCE OF LOWER-CHEST INFECTIONS IN
INFANTS IN GREAT BRITAIN

This shows the percentage occurrence of lower-chest infections as
reported in the study by Douglas and Waller (reference 32) over a
three-year period. There is an approximately threefold increase in the
occurrence of such illnesses as between infants being brought up in
moderate- and high-pollution zones compared with those in low-pollu-
tion regions. In table 7 are shown criteria of the air-pollution levels as
classified in this figure.

is clear that one would, on the basis of this evidence, expect
some increment of lower-chest infection to be occurring
amongst the infant age group in many North American cities
as a consequence of air pollution. The study conducted in
Britain has yet to be repeated on anything like the same scale
in the United States or Canada, and at the moment one has no
indication whether or not the results would be similar. One

might also infer that on the basis of these data it might be possible in a major urban area to demonstrate some variation in occurrence of bronchiolitis or lower-respiratory infection in infants in relation to the SO_2 or the particulate pollution of the region of the city in which they are being brought up. No doubt some studies of this kind will soon be reported from North America.

More recently, the preliminary results of a study of respiratory-illness rate in a number of families living in a region of Chattanooga in the United States have been published.[11] One area of this city is in proximity to a factory making T.N.T., and has a relatively high oxide of nitrogen and a low particulate pollution level. In other areas these ratios were reversed. The study appears to demonstrate a higher incidence of bronchitis in both infants and school children in zones in which the NO_2 averaged, as a six-month mean value, more than 0.06 ppm. A similar increment was observable in the fathers of the families, and a second study in the same area essentially confirmed the first finding of an excess of acute respiratory illness amongst children living in higher NO_2 polluted regions. It was noted in chapter 2 that annual average levels of 0.06 ppm are exceeded in more than half of the American cities with a population from 50,000 to 500,000 and in 85 per cent of cities with populations over half-a-million. This appears to be the only study to this point which relates data concerning oxide of nitrogen pollution to any observed adverse health effects.

Although no very detailed studies of this kind have yet been reported for the Los Angeles region, it has been found that there were no demonstrable differences in a simple pulmonary-function test amongst school children from areas of different levels of oxidant air pollution. There have been reports of children complaining during school recess of some soreness in the chest on days of high oxidant air pollution, and this is an early symptom of ozone exposure,[16] but measurements of pulmonary function synchronous with such complaints or during such episodes have not, as far as I know, yet been reported. Nor have any other comparisons yet been made in relation to lower-respiratory illness morbidity amongst children in this area as opposed to other sections of the United States.

The demonstration, admittedly subject to confirmation, that existing levels of air pollution may be having an adverse effect on the respiratory health of children not exposed by virtue of their occupation to irritants, and not yet cigarette smokers, constitutes in my estimate a new and disturbing factor in the consideration of the air-quality criteria at which the modern city ought to aim. The actual mortality produced, even in the worst regions, in this age group by air pollution must be extremely small, yet one has to be concerned at possible longer term effects. The work which has now been published on the problem will without doubt stimulate other studies in other regions of the world, but until data have appeared which suggest either different answers or an alternative explanation of existing findings, one has to accept that levels of air pollution seem to be associated with higher levels of respiratory disease in this important age group.

Lung Cancer

Although the association of cancer with compounds derived from coal tar has been known for many years, and data collected in Great Britain twenty years ago suggested that there were major differences in the incidence of lung cancer between urban and rural areas, it is now realized that the problem is much more complex. Patterns of cigarette smoking changed more quickly in urban areas than in rural ones, and studies which do not take full account of cigarette-smoking consumption are unreliable when it comes to making judgements about the possible role of air pollution. There are other reasons for looking closely at potential relationships between air pollution and lung cancer. Some polycyclic hydrocarbons such as benzpyrene can produce cancers in experimental animals, and these substances are emitted from diesel and gasoline engines and from some industrial processes. However, the level of concentration in city air is exceedingly low and there does not seem to be any considerable increase in lung cancer in men who, by virtue of their occupation, are exposed to levels of those concentrations hundreds of times greater than the general population encounters.

An additional difficulty in making comparisons between urban and rural areas is the very real possibility that the standard of diagnosis of lung cancer may be much higher in the city than in the rural district. A careful correlation of existing data, bearing these factors in mind, suggests that there is a slight incremental factor in the incidence of lung cancer related to urban living. There is a *greater* occurrence of lung cancer in rural Britain than in rural districts of the United States, and it seems to follow from this that pollution of the air in British towns is unlikely to account completely for the higher incidence of lung cancer in British urban districts in contrast to rural districts when occupation and smoking habits have been fully taken into account. Further complications in the analysis of these data have come from information from countries such as Norway which also show an urban excess of lung cancer, even though the general levels of air pollution in that country are well below those reported from Britain.

The study of the occurrence of lung cancer in Los Angeles indicated that when smoking and residential histories were taken into account the mortality was no higher than in other cities of the United States. There is some experimental evidence suggesting that in animals there may be interactions of factors such as sulphur dioxide, cigarette smoke, and benzpyrene in causing higher levels of lung cancer, and it is extremely likely that these factors are additive in terms of carcinogenesis, just as some of them probably are in a disease such as chronic bronchitis.

The lung-cancer death rate in men per 100,000 population has increased in Canada from 13.8 to 32.9 between 1950 and 1966. On the basis of all the evidence, and the overwhelming mass of data associating lung cancer with heavy cigarette smoking, it is fairly safe to say that all or almost all of this increased burden of serious disease must have been occasioned by cigarette smoking. At the most, one might attribute a 10 or 15 per cent incremental factor to modern urban living over and above the influence of cigarettes, but a great deal more work will be needed before any estimate of this kind can be considered anything other than a guess.

It is of great interest that people who have emigrated from

Great Britain to other countries, particularly South Africa, have been found to have an incidence of lung cancer higher than that in the countries they went to, in relation to their cigarette smoking, but lower than that of the population of similar age who stayed in the United Kingdom. There is no satisfactory explanation for this observation.

EFFECTS OF CARBON MONOXIDE

Carbon monoxide has very unusual properties in relation to living organisms, and because of this specificity this particular air pollutant deserves detailed consideration. In mammals the blood has special properties which enable it to carry far more oxygen than could be borne by a simple fluid. This is achieved by the presence in the red blood cells of a pigment, hemoglobin, which has a considerable affinity for oxygen, and to which oxygen can be linked. By virtue of this property of hemoglobin, blood can carry approximately *a hundred times*

TABLE 8

EFFECTS OF EXPOSURE TO CARBON MONOXIDE
FOR TWO HOURS

CO CONCENTRATION IN AIR ppm	SEDENTARY SUBJECT % COHb IN BLOOD	LIGHT EXERCISE (Ventilation doubled) % COHb IN BLOOD
20	1.25	2.5
40	2.0	4.0
70	3.8	7.6
100	5.0	10.0
200	10.0	20.0

The data have been taken from reference 19 and other published information. Transient CO concentrations of 40 ppm are not uncommon in major urban areas; in the seated subject a two-hour exposure to this level of carbon monoxide would increase the carboxyhemoglobin in the blood by 2 per cent. Heavy cigarette smokers, as a consequence of inhaling cigarette smoke, are not uncommonly found to have percentage of COHb levels as high as 6 per cent.

more oxygen than could be transported by an equivalent volume of water.

Carbon monoxide (CO) has the ability to combine with hemoglobin in much the same way as does oxygen, and, once combined, the hemoglobin is unable to fix an atom of oxygen since its position is occupied by a molecule of CO. This in itself would not be particularly dangerous were it not for the fact that hemoglobin has 250 times the affinity for CO than it has for oxygen.[19]

In air-pollution terms, there is little likelihood that the gas would be breathed for a long enough period for equilibrium to be reached. For this reason, in table 8, the calculated value of carbon monoxide in the blood is given under conditions of rest and light work as a consequence of two hours of exposure to different concentrations of the orders of magnitude that may be encountered in cities (see chapter 2).

By far the most important source of CO to which individuals are currently exposed is cigarettes. A person smoking more than twenty cigarettes a day, and who inhales the smoke, will be found to have a level of carbon monoxide in the blood between 3 and 6 per cent carboxyhemoglobin (COHb), and a very heavy smoker may be found to have as much as 10 per cent COHb. This would only be reached by a non-smoker after two hours of exposure to 200 ppm of carbon monoxide, a level rarely reached in city air and one to which a citizen is unlikely to be exposed under the most adverse circumstances for more than a few minutes. Carbon monoxide, once fixed to hemoglobin in the form of carboxyhemoglobin, is only slowly removed by ventilation from the blood and it takes approximately twelve hours of non-exposure for the blood carbon monoxide to return to half its pre-existing level. In this sense it is cumulative, but a period of 24 or 36 hours of non-exposure will result in most of the carboxyhemoglobin having been removed.

Because carbon monoxide has been known to be a potent poison and an important hazard for nearly one hundred years, a great deal of experimental work has been done on it. Whenever anything is combusted incompletely in a closed space there is danger of release of carbon monoxide, and many fatalities have occurred because ventilation of a tent or a cabin or a motor car was inadequate in the presence of CO being generated. Less work has been done on the effects of very low concentrations of carbon monoxide, but in relation to air pollution it is these effects which are important.

Slight differences in perception by normal subjects have been used to test at which level of carboxyhemoglobin some measurable impairment of appreciation of relative brightness, or visual acuity in general, becomes affected. The acute elevation of carboxyhemoglobin from 0 to 10 per cent COHb shows that this level of CO will impair many aspects of visual and auditory discrimination.[9] It will also cause an elevation of pulse rate. Some tests, particularly those involving discrimination between two auditory tones, and sensitive tests of visual threshold, show effects when COHb levels are below 5 per cent, though these experiments are very difficult to control. These results are obtained by producing step increases in COHb, and it is not at all clear whether, if the COHb level is always approximately 5 per cent in a cigarette smoker, any continuous decrement in these aspects of performance is necessarily present.

Overt and evident symptoms of carbon monoxide intoxication are unlikely to be noticed until the concentration in the blood rises to between 10 and 20 per cent COHb. One of the most prominent early symptoms is headache, but by the time this has become evident the deficiencies in psychomotor performance are easily measurable.

From the very considerable body of experimental work on human subjects on which these general conclusions are based,[9, 19] it may be concluded that exposure to the levels of carbon monoxide that now commonly exist in cities is unlikely to lead to increases in COHb, under the most unfavorable circumstances, of more than approximately 5 per cent. These levels are those at which just measurable changes are produced in acute experiments, but it seems unlikely that they would adversely affect to any considerable extent the capability of a motorist to drive his car or his judgement in relation to driving situations. There will be special circumstances occurring in traffic jams, underground-parking garages, tunnels, and possibly when a heater intake is positioned so that the exhaust from the car in front may be delivered into the passenger compartment in which higher amounts of COHb may well occur.

Until recently, it was not believed that levels of carbon monoxide below 5 per cent had any significant long-term effects on the body as a whole. Very high concentrations were known to be capable of producing damage to the central nervous system, but no ill effects had been attributed to low concentrations. During the past few years, workers in Copenhagen in a series of important animal experiments have produced evidence that low levels of carbon monoxide over a longer period of time may affect the permeability of small blood vessels.[19] Although the relevance of these animal experiments to the human situation is bound to be subject to continuing discussion, the demonstration that low levels of carbon monoxide (combined with a high fat diet) do produce evident changes in the blood vessels of the animals is an important and disturbing observation. Levels of 11 per cent COHb in rabbits together with a high fat diet have been shown to produce changes in the tissues of larger and medium sized blood vessels that are similar to those thought to occur in human arteriosclerosis. The primary change is believed to be increased permeability of the vessel wall to one of the proteins in blood, serum albumin.

Although it has long been known that cigarette smoking enhances the risk of developing coronary artery disease, there have been up to this time very few ideas as to how this interaction might occur. The relationship between cigarettes and lung cancer could at least be easily visualized, but it was far from clear how cigarette smoking could be having an adverse effect on the blood vessels supplying the heart. It is not yet established that carbon monoxide is in fact the link between cigarette smoking and heart disease, but at least it is a potential link. The recent work on the effects of carbon monoxide, chronically in the blood, on small and medium sized arteries is disturbing because it does provide a link for a relationship which has been proved consistent in a large number of epidemiological surveys. It is too early to say that chronic COHb levels of 5 per cent are hazardous in the long run, but it is also unwise to assume that low concentrations are without long-term effects. It is to be expected that further work will throw more light on the role of CO in relation to coronary heart disease but, until this work has been done, it would be wise to reserve judgement on the possible responsibility of these low levels of carbon monoxide as an etiologic factor in cardio-vascular disease.

A very recent study of daily mortality in Los Angeles County appears to have shown that there is a significant association between community carbon monoxide concentrations and mortality.[49] This kind of evidence, if it can be confirmed from other cities, deserves close study since ambient levels of carbon monoxide may possibly throw a burden on an individual already suffering from advanced coronary artery disease or some other conditions. By reducing the amount of oxygen the blood can carry, a 2 or 3 per cent increase of COHb when associated particularly with some other stress, such as a day of exceptionally high temperature, might possibly be sufficient to produce a coronary thrombosis.

MISCELLANEOUS EFFECTS

A number of miscellaneous observations concerning human health and air pollution have not been covered in the preceding sections. In industrial areas of Great Britain, it has been shown that chronic infection of the middle ear (chronic otitis media) is commoner in regions of high pollution; this may be a separate factor or be related to the higher incidence of respiratory illness in children which has been documented to occur in many of these cities. In Czechoslovakia, there has been concern about the effect of *nitrogen oxides* and of *sulphur dioxide* on the blood of children living in areas of high pollution as shown by the level of methemoglobin in these children. This compound might be produced by high levels of atmospheric oxides of nitrogen, but the significance of the changes in the blood is hard to interpret.[11] Some evidence of anemia and even rickets has been reported from the USSR amongst children living in a highly polluted area, but the details in the publications are insufficient to enable a precise evaluation of the significance of the study.

It has been demonstrated that slightly higher levels of *lead* exist in the blood of individuals living close to major highways. These levels do not in themselves constitute a risk, but the fact that they can be demonstrated at all indicates the long-term hazard of this particular pollutant. It has also been reported that some animals in metropolitan zoos have been affected by excessive lead;[18] although it has been assumed that the source is atmospheric lead from automobiles, this does not seem yet to have been precisely established. Similarly, reports of blood changes in children from Czechoslovakia have not been closely related to blood levels of lead and their origin remains obscure.

The *photochemical-oxidant* pollution, by the production of peroxyacetyl nitrate (PAN) compounds, leads to eye irritation and soreness of a transient nature. Although this discomfort is very generally experienced, it is not thought to have any permanent ill effects on the eyes. As noted earlier in association with some industrial plants which cause emission of *hydrogen fluoride*, some secondary effects might occur in humans as a result of eating vegetation with a high fluorine content, or drinking milk from cattle which have grazed on pasture with a higher than average fluorine content. No direct human-health effects have been demonstrated by this mechanism, though it is not inconceivable that some may have occurred.

Asbestos has long been known to be a hazard to the lungs of man, and those working with this material at a time when elementary precautions against inhaling the fibres were not taken suffered a considerable incidence of fibrosis of the lungs, often with the later development of lung cancer. During the last ten years asbestos has come to be used in a very wide range of secondary industries and, as a result of this, many more workers and individuals are exposed to it than in the past. It has been incorporated into paint, used for the manufacture of ceiling acoustic tiles, used for undercoating automobiles, and most recently has been sprayed onto steel work during the construction of buildings. The tonnage of asbestos being mined in Canada has increased from about 400,000 short tons per year in 1940 to over 14,500,000 short tons in 1967.

A number of studies, including one completed in Montreal, have shown that individual asbestos bodies may be detected during autopsy in the lungs of people who have not been exposed to asbestos through their occupation. There appears little doubt that these bodies are in fact asbestos fibres, and one or more may be found in the lungs of about half the normal population of cities such as Montreal, London, Sydney, or Cape Town. A single asbestos fibre has been trapped from the air of the city of London and photographed with an electron microscope. The lung has no ability to dispose of these particles which become walled with a protein-rich material and remain in the lung for very long periods of time. It is not to be expected that a small number of these fibres exert any adverse influence on the health of the individual, though there is concern that the uncontrolled secondary use of asbestos is in fact leading to higher levels than in the past.

It is not at all clear exactly where the fibres which may be found in the lungs of a modern city dweller have come from. Initially, it was thought that perhaps they had been burnt off the brake linings of automobiles, but a more recent opinion is that it is unlikely they could remain intact in a recognizable form from this source. It is possible that construction work in general leads to the wind dispersal of significant amounts of asbestos, and the risk of dispersion is obviously very much greater as a result of the recent technique of spraying asbestos on parts of buildings as these are being constructed. It seems clear, however, that as a general air-pollution problem asbestos cannot at the moment be considered a major danger.

There is no evidence to suggest that one or two asbestos bodies in the human lung have any adverse effect of themselves, and even among those in the asbestos industry very high levels of exposure over many years must have occurred before any recognizable ill effects are demonstrable. Indeed, in the industry as a whole, the improved engineering associated with ventilation and the automation of processes previously done by hand has meant that exposure levels are very much lower than they were in the past and adverse effects are consequently much less common. However, this is not to say that continuing review of the use of asbestos is not required, and it will be argued later (see chapter 6) that consideration should be given to methods of controlling the uses of asbestos in the open air where these are likely to result in some general air-pollution hazard.

In the concentrations that occur in cities, it is very unlikely that oxides of nitrogen would have any significant acute effects on human beings. However, the concentrations of *ozone* which are being approached (which may reach 0.6 ppm) are getting close to or may actually have reached quantities capable over a two-hour period of producing measurable changes in human pulmonary function.[16]

Ozone is an intensely irritant gas, and light exercise in an environment of 0.7 ppm of ozone produces evident irritation of the windpipe of which the subject is conscious, and also leads to a considerable increase in the resistance to airflow in the human bronchial tree. Thus, although oxidants have not been shown in Los Angeles to be associated with any measurable increased mortality, there seems little doubt that in levels being approached in Los Angeles County and possibly elsewhere they are capable of producing acute irritation of the respiratory tract sufficient to cause a measurable change in function. A recent report in the newspapers indicated that a severe photochemical smog episode occurred in Rotterdam in October 1971, but exact details of this have not been published. Data from Tokyo indicate that a sequence of events similar to that noted in figure 3 also takes place in that city.

HEALTH COSTS OF AIR POLLUTION

In the light of the many uncertainties inherent in the existing data, it might be considered foolhardy or premature to attempt any estimate in terms of dollars of the present cost of air pollution in relation to human health. An effort to arrive at some provisional estimate of these costs has recently been made by two economists whose paper should be consulted in the original by anyone interested in this formidable problem.[58] Having reviewed most of the existing data, and coming to some preliminary estimates on the basis of regression relationships between respiratory illnesses and levels of air pollution, they conclude that even after attempting to be as conservative as possible: "The total annual cost that would be saved by a 50 per cent reduction in air pollution levels in major urban areas (in the U.S.A.), in terms of decreased morbidity and mortality, to be $2080 million." They point out that a more reliable figure is probably that a "50 per cent reduction in air pollution in major urban areas would result in a 4.5 per cent reduction of all economic costs associated with morbidity and mortality." They point out that this latter figure is not sensitive to the exact figures chosen for calculating the economic costs of ill health.

A number of other important points are made in their study, which the authors feel probably underestimates the relevant costs. They point out that until such time as something other than pollution is shown to be the agent responsible for an "urban factor" in the existing data, it is fair to conclude that the urban factor being studied is, in fact, consequent upon levels of air pollution. There is often a tendency to expect that a decisive component remains to be described which has not so far been measured, and these authors insist that, until such time as this magic factor is produced and studied and substantiated, those studies which show urban/rural differences after all other variables have been taken out must be assumed to be reflecting differences related to air pollution. In terms of lung cancer, they have taken a ratio of approximately 34 per 100,000 for rural areas as compared with 56 per 100,000 in cities with a population of over 50,000.

The kind of analysis that they have attempted may have to be modified in the light of more recent information, but it may be taken as the best estimate that can be produced from the existing data, and it may well turn out to be conservative. On a population basis, it may be calculated that the health costs of air pollution might well be of the order of $200 million per annum in Canada. The current rate of chronic bronchitis and lung cancer in Canadian cities seems to be very similar to that in American cities, and on a straight tenfold population difference one might guess that Canadians should assess the health costs of air pollution as of this order of magnitude. Much of this outlay is related to loss of working time by diseases which are aggravated by air pollution, and the remainder is mostly incurred by hospitalization charges and medical diagnostic expenses involved in the diseases themselves. However one looks at these figures, the aggregate costs are substantial.

SUMMARY

It is not easy to summarize the enormous amount of work which has been done in the last ten years on the relationship between human health and air pollution without leading to generalizations that are so broad as to be liable to be misleading, on the one hand, or without getting into such a mass of technical detail that the reader quickly loses sight of the essentials. Keeping track of the papers relating air pollution to respiratory disease alone can easily become a full-time job, but it is encouraging that an increasing amount of study has been devoted to this important problem. From many parts of the world the same kind of relationships have emerged, and it is this reinforcing of data with support from different studies by different investigators, amongst different people, which lends a great deal of strength to the main case.

It is still sometimes said that air pollution has not been shown to cause any disease. In part, this confusion of thought arises from an attitude which would expect that some disease, *absolutely distinguishable from all other diseases*, might be expected to be attributable to air pollution, but this is not so. If one compares a pediatrician in an industrial region with his counterpart practicing in a rural area, it will be clear that the urban practitioner will probably see more cases per head of population of lower-chest infection or bronchiolitis than his rural counterpart. But it is very important to recognize that the threefold increase in incidence in these conditions documented in Great Britain between rural and urban areas (see figure 7) does not mean that the increased number of cases being seen in the urban region represents some disease clinically distinct from that seen in the country. The conditions appear identical, and yet there seems little doubt that the increased numbers being seen in the urban centre are associated with the air pollution in that region. Similarly, a farmer who smokes thirty cigarettes a day for twenty years may have chronic bronchitis that looks the same as chronic bronchitis seen in a bank manager with a similar smoking history in a city. It is erroneous to conclude that the bank manager's chronic bronchitis would not have occurred if he had lived in the country, though it may be safe to say that had the farmer lived in an urban environment he would have had more episodes of acute chest infection; at equivalent age his pulmonary function might be found to be more impaired, and he might suffer from more cough and sputum as a result of his urban living.

These are difficult points of emphasis to understand, but they must be stressed to offset the tendency of news media to try and seize on a single statistic or piece of information and enlarge it in a dramatic and falsifying way. Any physician who states that a *major* burden of lung cancer or infant *mortality* is to be associated with air pollution is guaranteed a headline in the newspaper, an interview on television, and an opportunity to speak widely in public, although the scientific basis for his statements may be totally lacking.

It is noticeable that the general ecological effects of air pollution on plant life and animals, which have become rightly a subject of major international concern, have tended to outweigh the detailed consideration of the existent direct health effects on humans. For every physician concerned about the problem of growing numbers of Canadians living in cities, and of increasing levels of air pollution in these cities, there are maybe fifty biologists whose interest relates to the conservation of wildlife and the possible long-term effects of detergents and pesticides on the countryside and the animals it supports. This has had the effect of diverting attention away from human problems, and deserves some correction.

It will be argued in a later section that the long-term control of air pollution is a complex matter involving most segments of society. The present data on the effects of urban atmospheres on human health seem to justify the view that steps towards control of urban air pollution should be taken immediately, since further deterioration in large cities would almost certainly be associated with greater effects than those which are probably already occurring.

The question is often asked whether an improvement in health has been documented to occur in a city where air-pollution levels have been lowered. As noted above, the reduction in sulphur dioxide and particulate pollution in London has been followed by a lowering of mortality in relation to peak levels of smoke—now very much lower than was customary twenty years ago.

Very recently an important study has been published of the re-survey of a New Hampshire town with a pulp-mill industry. The town was surveyed in detail in 1961, with a repeat in 1967. During this interval, the level of SO_2 pollution had been reduced by about half and there had been a similar decline in dustfall. In spite of the fact that the 1961 levels of these pollutants had not been very high, the study revealed a reduction in the prevalence of respiratory symptoms in all age groups, and a small but definite improvement in the performance of pulmonary-function tests. The authors concluded: "It is believed that the decrease in air pollution could account for the decreased prevalence of disease and the slight improvement in pulmonary function" (see Ferris, B. G., Jr., *et al.*, American Review of Respiratory Disease, Volume 104, 232–244, 1971).

chapter five

LONG-TERM PROBLEMS AND NON-PROBLEMS

INTRODUCTION

The social historian twenty years from now may be able to define fairly precisely the forces within western society which came together in the latter part of the 1960s and resulted in a surge of public concern relating to environmental quality. It was not that a substantial amount of new information became available, since much of it had been known for years; nor was it some sudden increase of concern occasioned by a massive disaster such as the London smog episode of 1952. Amongst the prominent factors, however, were the recognition that species of birds were being decimated by residual DDT as a result of biological processes which, up to that time, had not even been suspected, and the general fear that western man was set on a course which could only end in disaster. The fact that the United States of America was using natural resources at an ever-accelerating rate, and that the poorer countries of the world had still to control their population growth, became matters of public discussion. Everyone suddenly became aware that decisions were being taken concerning exploitation of natural resources, or to follow some technological advance, without any consideration of the long-term consequences of such actions.

It needed only the juxtaposition of the data relating to exploitation of natural resources and increasing energy consumption, on the one hand, and an inventory of where a continuation of such policies would lead, on the other, for everyone to see that the unquestioning pursuit of increasing material prosperity in the short term would in the long term constitute a disastrous national and international policy. It became apparent that, ever since the beginning of the industrial revolution, industrial decisions had been taken with a reversed order of priorities. Possibly the last such undertaking which will be in this category (and in the general tradition of all such development up to that time) was the development of supersonic passenger aircraft. After the feasibility of passenger flight at supersonic speeds became evident, the decision to build a prototype aircraft was taken by the governments of Great Britain and France together, and the project began. After the prototype had flown and been tested, and many millions of dollars expended on the program, it became known that it would be necessary to make and sell at least 200 of these aircraft for the commercial venture to break even.

At this point the question was asked as to what would be the long-term effects, produced by a fleet of such aircraft, of deposition of water vapour and particles at altitudes of 60,000 feet. As noted below, a preliminary attempt was made in 1970 to get an answer to that question. In retrospect, one might remark that it would have been better to have asked the question first, and spent one or two million dollars trying to arrive at the best estimate of the consequences of aircraft flying in large numbers around the earth at those altitudes. Once it had been agreed, *if ever*, that from the ecological point of view this was feasible, development of a prototype might have been begun. Thus, the Concorde project represents a series of decisions taken essentially backwards. It seems very likely that it may be the last major technological decision taken in that way, and that in future the ecological and environmental implications will be studied in advance of the development.

No discussion of long-term problems associated with air pollution would be complete without a word relating the problem of air pollution to wider issues of environmental concern. Many of these hinge on the complex interrelationship between growth of population, growth of consumption, and increasing industrial and technological development. Many people, particularly those with a puritan background, are quick to attribute man's present problems to inherent greed or love of luxury. As has been pointed out, however, this is not so important as is a decision-making process which may favour one industrial product over another, and by its success leads to the general acceptance of a product which either uses more energy or produces more pollution. Many examples from ordinary life may be taken to illustrate this.

The development of the motor car in the late 1930s, and for twenty years afterwards, led to the use of higher compression automobile engines with a shorter stroke and a much faster operating speed. These engines were more powerful in relation to their size and also more economical in terms of gasoline consumption. However, they produce more pollutants in proportion to their fuel consumption than did older automobile engines. The technological changes were not accompanied by any consideration, during those years, of the effects the change would have on the environmental quality of the cities.

The development of man-made textiles and the fact that these are longer wearing and have some other desirable properties in relation to cotton have meant a general acceptance of them. However, the energy needed to make cotton comes from the sun and cotton fields produce no pollution. The energy to create man-made fibres comes from sources of fuel, and their manufacture is accompanied by emission of secondary products. Hundreds of such examples are partly responsible for the tremendous increase in energy needs per capita which has occurred in the United States and which it has been predicted will continue to rise.

It is of interest that the correlation between a nation's per capita use of energy and its level of economic development is almost linear. A plot between gross national product per capita in dollars and energy expenditure per capita in millions of thermal units shows that at one end of the scale is the United States, in which the gross national product per capita is estimated at approximately twenty-eight hundred dollars and the energy per capita as about 160 million thermal units, and at the other end of the scale is a country such as Brazil in which the gross national product per capita is approximately two-hundred dollars and the energy expenditure per capita is roughly 15 million thermal units.[20] Countries such as Japan, France, USSR, United Kingdom, and Canada scatter fairly uniformly about the line that joins these two countries. Energy consumption in the United States, plotted in quadrillion units of heat, is believed to have increased from approximately 15 in 1910, to 25 in 1940, and to more than 60 in 1970. At the rate in which the change is occurring, the figure will exceed 120 by the year 2000. Since 1940, the increased energy consumption has been largely attributable, on a percentage basis, to the increasing use of oil and natural gas.

In this volume no detailed account can be given of the complex problem of the energy needs of modern society, but it has become apparent that the appetite for energy will outstrip our potential to satisfy it. The generation of electricity from nuclear sources poses major problems in the long-term disposal of the radioactive material. Generation of electricity from hydroelectric power, which is virtually pollution-free, is not limitless and, indeed, many of the existing sources are considered to have been tapped already.

The resource of natural gas, which has been developed only in the last ten years, is far from infinite and, although this fuel offers virtually pollution-free heating in the urban environment, the opportunity to make much use of it will apparently not last more than twenty or thirty years. Eighty-five per cent of the electricity in the United States is still generated from coal, and if this supply is to meet the projected energy needs of that country, and if precautions against sulphur emission were to be unchanged from what they were a few years ago, the tonnage of sulphur dioxide released from this source alone would double between 1960 and 1980.

Although increasing population of itself is one aspect of the pollution problem, it has been the growing energy utilization per capita, for a variety of different reasons, which has been mainly responsible for the increasing release of pollutants.

The mistake is often made of assuming that air pollution and its problems are exclusively the difficulties of modern industrial man, and will remain so. It is true that the problems have arisen first, to any extent, in relation to dense urban conglomerations of people with uncontrolled pollutant emissions; but it is not true that they will remain for very long the exclusive concern of such societies. One has only to contemplate the industrialization of China to realize that global pollution problems would become severe if presently underdeveloped nations of the world had a pollutant emission per capita similar to that of the western world.

In some cities in the Middle East the automobile density is now approaching that of cities in the West, and because of the bright sunlight in those regions, and the often static meteorological conditions, photochemical smog is already becoming a problem, as any perceptive tourist may note. Obviously an energy consumption per capita as it now exists in the United States could not conceivably be extrapolated to cover the whole world, since the known energy resources could not meet such a demand. It is also quite evident that the pollution emissions the western world has tolerated could not be extrapolated to presently underdeveloped countries without producing major problems for the global environment.

In the ferment of public discussion on air pollution which has occurred over the past three or four years, it has been extremely difficult to keep much sense of perspective as to where the priorities lie. There has been, furthermore, a lot of uncritical publicity and speculation concerning different aspects of the long-term problems associated with air pollution, and public anxiety over the question has been reflected by a good deal of emotion concerning issues that appear to be essentially "non-problems." In the two sections that follow, the long-term implications of air pollution are considered in the light of whether these appear to be at the present time "non-problems" or problems and then to assist the reader to distinguish between these issues.

NON-PROBLEMS

Oxygen Depletion

When I was taking part in an open-line radio program on environmental problems, the first question to come in from a member of the public was whether the panel believed that the oxygen in the air would run out in twenty years or sooner. A very careful calculation by Broecker[22] has shown that if, in the course of a few days, mankind were to burn up all the known fossil-fuel reserves, we would use less than 3 per cent of the available oxygen in the atmosphere. His computations show that a general depletion of the atmospheric oxygen supply by burning fossil fuel is not a possibility in the foreseeable future. Broecker has also calculated that our total combustion processes to this point of time might have used up about 7 of every 10,000 oxygen molecules available to us. It is clear from his figures that, although the high oxygen demand by organic and inorganic material we have added to our fresh-water lakes and rivers has in many instances reduced the level of oxygen in these waters below that required by fish, the oceans still represent an enormous oxygen reservoir in the earth's environment.

The oxygen cycle of the biosphere is complicated by the fact that oxygen appears in very many chemical forms and combinations.[20] It is now believed that the oxygen in the earth's atmosphere accumulated primarily as a result of photosynthesis by plant life, and that free oxygen began to build up in the atmosphere approximately 1.8 billion years ago. The oxygen in the atmosphere is thought to have approached its present level approximately twenty million years ago.

It may be concluded that depletion of oxygen by the combustion of fuel is not a problem. Interference with marine photosynthesis by pollutants in the earth's atmosphere should perhaps be a matter of concern, but a discussion of this problem is beyond the scope of the present volume.

Carbon Monoxide Accumulation

Although the tonnages of carbon monoxide being added to the atmosphere are far from negligible, and have been calculated to be in excess of 250 million tons per year,[9,19] the change in atmospheric background level of carbon monoxide has been much less than would have occurred if it were remaining in the atmosphere. If there were no mechanism for the conversion of CO to something else, the average global background level would by now have exceeded 1 ppm. However, it is known from measurements that the background levels have not risen, but have remained relatively constant. At this point of time, no final opinion can be given on what is responsible for the removal of CO from the atmosphere. Early suggestions that this might be occurring in the oceans have not received support from recent scientific studies that indicated rather that warm seas act as a source of carbon monoxide, probably as a result of decaying vegetation.

Some very recent experiments[54] have suggested that the soil contains a large number of bacteria which are capable of removing carbon monoxide from the air above the soil, and it is of considerable interest that the capability of soils to do this varies considerably. The scientists investigating this problem have shown that a square metre of soil from the coastal redwood districts of California, for example, is apparently capable of taking up approximately 61 mg per hour of carbon monoxide.[54] By comparison, other samples of soil can only take up a third or a fifth of this amount. The experiments that have been reported appear to show clearly that the ability of soil to take up carbon monoxide is destroyed when the soil is sterilized, and it is assumed that the phenomenon is caused by the activity of microorganisms. An approximate estimate of the soil's faculty to remove carbon monoxide from the atmosphere indicates that the capacity of the total soil surface of the continental United States to take up CO is about six and one-half times greater than the annual estimated production of carbon monoxide attributed to the United States. The authors of this study conclude that "the soil, therefore, must now be considered as a major natural sink for CO that is released into the atmosphere by natural emitters or by the burning of fossil fuels."[54]

On the basis of this calculation, and since atmospheric carbon monoxide is known not to have increased over the past thirty or forty years despite enormous emissions of this gas, it seems fair to regard the accumulation of CO in the earth's atmosphere as not a problem in the foreseeable future.

Sulphate Excess

The burning of fossil fuels containing sulphur inevitably results in the dispersion of this element as sulphur dioxide or sulphates into the earth's atmosphere. As noted earlier (see chapter 2), the tonnage per annum of sulphur dioxide so produced is far from negligible. Sulphur and sulphates are intrinsic to living matter since sulphur provides an important part of the structure of proteins. Sulphur is therefore a necessary constituent of animal body protein, and recently much attention has been given to the global problems of recycling and storage of sulphates.[20] Although many parts of this story remain to be filled in, it seems clear that bacteria which are capable of reducing sulphates to the form of hydrogen sulfide (H_2S) play a major part in these natural cycles. At the present time the release of sulphur dioxide as a consequence of man's activities does not appear to be of sufficient magnitude to dislocate these relationships. The same cannot be said of phosphates and phosphorus, but since these are not air pollutants the problem cannot be considered further in this volume.

In certain regions the emission of very large amounts of sulphur dioxide may possibly produce undesirable long-term effects in the downwind area. An increase in the acidity of rainwater has been reported in some areas of Holland and Sweden and this is believed to be a consequence of the major industrial activity concentrated in Western Europe. The increased acidity in small lakes and rivers is thought to be at a point where the biological ecosystems, particularly certain types of fish, may be adversely affected by this alteration. The fact that long-term sulphate emission may be regarded at the present time as not a major concern does not mean, therefore, that the consequences of SO_2 emission in certain regions are not matters of proper anxiety.

PROBLEMS

Nitrogen Fixation

During the past few years there has been increasing concern that the nitrate balance of the earth might be adversely affected by human activities. The extensive use of manufactured nitrates as fertilizers and the increasing run-off of these into rivers and fresh-water systems, together with the deposition of nitrogen-containing raw sewage into the same bodies of water, over the course of time clearly influenced adversely the natural nitrogen cycles. The total ecological nitrogen balance is, in quantitative terms, not completely understood, although the broad outlines are clear. The demand for explosives in the twentieth century in fact provided the major incentive for the invention of an industrial process enabling atmospheric nitrogen to be fixed.

The nitrogen-fixing bacteria present in soil and in relation to plants can accomplish at ordinary temperatures and pressures what takes a considerable amount of energy in a factory, and the mechanisms whereby the bacteria complete the fixation are not yet fully understood. Various crops, particularly legumes, represent a very large natural source of fixed nitrogen, and until recent years these natural cycles far exceeded human contributions. Electrical storms have the effect of fixing nitrogen and of forming oxides of nitrogen in the air, and the amount of fixed nitrogen delivered to the earth by rainfall has been calculated in the order of 25 million metric tons per year. About a third of this may be in the form of atmospheric nitrogen fixed by lightning.

However, the actual quantity of nitrogen being fixed (in the sense of being discharged into the atmosphere) by industrial processes or automobiles as oxides of nitrogen is relatively small in relation to the large amounts of nitrates being contributed as fertilizers. For this reason, nitrates regarded solely from the point of view of air pollution are such a small fraction of the total that they cannot be considered to comprise a major problem. It is very important to keep this distinction in mind, since the total nitrate production, and particularly the increasing nitrate of fresh water, is undoubtedly a matter for proper contemporary concern and continuing study. The most important effect of oxides of nitrogen in the atmosphere is either to act as primary irritants or to be the precursors of photochemical oxidants.

I have been present at a panel discussion on air pollution at which an industrial representative dismissed the contribution of automobiles, in terms of oxides of nitrogen, by virtue of the fact that the total nitrate burden this represented was but a small fraction of the total—a point of view which completely ignores the major distinction between liberating oxides of nitrogen as gases into the urban atmosphere, on the one hand, or spreading a tonnage of nitrates on the fields, on the other. In the long term, therefore, one may say that the nitrates contributed as air pollutants cannot be considered a continuing problem, though this is not to say that the total nitrates going into the biosphere, particularly from fertilizers, do not represent an important future hazard by upsetting the nitrogen levels in a variety of ecosystems.

Increase of Global Carbon Dioxide (CO_2)

Whenever a fuel is completely combusted in the presence of oxygen, carbon dioxide is evolved. The human body is an energy machine consuming oxygen and, by burning sugars and other foods, generating carbon dioxide which is expired. At the present time, between five and six billion tons of "fossil" carbon dioxide per year are estimated to be released into the atmosphere, and it has been calculated that this will be enough to increase the amount in the air by 2.3 ppm per year if all the molecules released were uniformly distributed and there were no mechanism of removal. A summary of scientific data concerning atmospheric carbon dioxide suggests that the CO_2 content of the atmosphere has risen from about 290 ppm to 320 ppm within the past century, with more than a fifth of the rise occurring in the past decade.[70] This increase would account for only about a third of the estimated carbon dioxide being released from all sources.

Although large quantities of carbon dioxide are known to be removed by the vegetation on earth and also by the disappearance of CO_2 into the ocean reservoir, the acceleration in fuel consumption implies that the amount of CO_2 in the atmosphere may climb from its present value of 320 ppm to very nearly 400 ppm by the end of the century. And there is justifiable concern over the long-term effects of such a change on the atmosphere of the earth. The calculation of the effects on the earth's climate of increasing CO_2 is not simple. It seems clear that an increase in CO_2 in the earth's atmosphere will reduce the outgoing infrared energy flux to space. This will cause a rise in the earth's temperature. But an exact tally of what increase in temperature would occur, on the basis of any given increase in global CO_2, involves assumptions about a number of variables in the earth's climate which are imperfectly understood.

The most recent computation[70] concludes that an increase in the amount of CO_2 in the earth's atmosphere by a factor of eight times its present level would produce a rise in surface temperature of less than 2°C. Taking 1971 levels for CO_2, the increase from 320 ppm to 400 ppm which may occur between now and the year 2000 would be predicted to produce an increase of approximately 0.1 of a degree Centigrade in surface temperature. Thus it appears that although there is some reason for concern about the increase of carbon dioxide in the global environment, and although it is clear that this should be very carefully watched over the coming years, there does not seem to be much possibility that CO_2 will produce a major change in the earth's climate within the next 30 years or so.

There have been some alarming projections in the past as to

the consequences of marked changes in the earth's temperature produced as a result of CO_2 accumulation, leading to melting of the ice caps of the earth and a significant rise in the level of the sea. It now seems likely that before any such stage is reached far more formidable problems relating, for example, to population control will have come to dominate the global scene, and there is little reason why the accumulation of CO_2 should be regarded as one of the major problems now confronting mankind.

Particulate Pollution

The effects of small particles accumulating in the earth's environment is, in a sense, the reverse of that which occurs as a consequence of CO_2 accumulation. Dust in the global atmosphere serves to reduce the surface temperature of the earth by virtue of the effect it has on sunlight. This has already been documented to be occurring over urban areas, and zones of high particulate pollution have been shown to have fewer hours of sunlight and a reduced sunlight intensity compared with rural areas only a few miles distant (noted in chapter 3). In 1968, a detailed comparison for this factor between the city of Montreal and a neighbouring region showed that during the period from 1 November 1965 to 31 May 1967 Montreal received on the average only 91 per cent of the total incoming solar radiation received at a site 15 miles northwest of the town centre. The attenuation was greater during the winter months. The elimination of a great deal of the smoke from the atmosphere of London has led to a 30 per cent increase in hours of sunlight over that city.

As has been noted in chapter 2, one of the features of the particulate pollution of cities has been that the particles have become smaller, since open-burning of coal now is either carefully controlled or not used at all, and the increasing automobile burden has meant a considerable emission of small particles. For the New York–New Jersey area in 1966 it was estimated that cars were producing 22,000 tons of particles per annum and diesel vehicles 11,000 tons.[12] One consequence of producing larger numbers of smaller particles is that these settle to earth very much more slowly than larger ones, and therefore may make a substantial contribution to the global particulate pollution.

The data on whether global air pollution has been increasing have been somewhat conflicting, but measurements of atmospheric turbidity in Antarctica in 1966 and a comparison with earlier data suggested that no major change in Antarctic turbidity had occurred between 1950 and 1966. A more recent report of careful observations made in Hawaii between 1956 and 1970[38] indicates that in this region there have been no major variations in atmospheric turbidity as a consequence of increasing particulate or aerosol pollution. The 1963 data from Hawaii show an interesting discontinuity which coincided with the eruption of a volcano (Mount Agung) in Bali. The eruption of Krakatoa in 1883 led to a great increase in dust encircling the earth, which had a measurable effect on the earth's temperature for a period of several years. After this fantastic eruption, the sound of which was noted four hours later three thousand miles away to be "like the roar of heavy guns," dust fell ten days later several thousand miles distant. The data from Hawaii indicate that there are annual cycles of atmospheric transmission of solar radiation, but there is no evidence of a general decrease in transmittance over the last fifteen years.

This is, however, a matter of legitimate concern, since it has been calculated that the earth's temperature is much more sensitive to a rise in global aerosol background, consisting of particles and droplets, than to an increase of carbon dioxide.[70] Growth by a factor of four in global aerosol background concentration would apparently be sufficient to reduce the surface temperature of the earth by as much as 3.5°C, and if sustained over a period of several years this would have a major impact on the earth's climate.

If these conclusions are correct, it follows that man's activities to this point of time have not led to any dangerous increase in total global particulate concentration. It is also evident that this aspect of pollution should be carefully monitored on an international collaborative basis, and the impact of changes, if such begin to occur, should be studied with the utmost care. The problem of an increase in aerosol concentration as a result of supersonic aircraft, a highly specialized question and somewhat distinct from particulate pollution generated at ground level, is dealt with in the next section.

SPECIAL PROBLEMS

Lead Accumulation

As noted earlier, if gasoline contains lead there are considerable emissions of this element in gasoline exhaust. There is no doubt that particulate lead concentration in some areas has been increasing at a spectacular rate; in San Diego, for example, a rise of approximately 5 per cent per annum has been documented and in other areas the increase has been greater than this.[24] There has recently been some concern that animals in metropolitan area zoos may be affected by atmospheric lead, although this conclusion has not yet been completely substantiated.[18] However, whether or not contemporary atmospheric levels of lead pose any *immediate* health threat to humans or animals is not the primary question relating to lead accumulation.

Lead emitted into the air in small particles ends up customarily in vegetation, and one way or another is very likely to find its way into food sources (milk, vegetables, and a wide variety of consumed items), in addition to the relatively small quantity likely to be absorbed into the body by direct deposition into the lungs, and clearance from there into the blood. There seems no doubt that the continued large-scale burning of gasoline containing lead would have the effect, over a fifteen- or twenty-year period, of steadily increasing the burden of environmental lead in a wide variety of ways. In 1967 the four U.S. producers of lead alkyls made 685 million pounds of tetraethyl and tetramethyl lead valued at about 254 million dollars. And it may certainly be questioned whether the global environment, broadly considered, can be allowed to accept such a burden indefinitely. A continued use of lead in gasoline on that scale would have the effect of slowly, *but inexorably,* increasing the lead in the general environment, particularly in areas close to heavy traffic density.

These long-term concerns provide the reason why many people desire to remove lead altogether from gasoline. That lead from automobiles is directly absorbed into the human body has been shown by studies which reveal higher blood concentrations of lead in people living close to main traffic arteries. There is no reason to suppose that these very low levels have had any deleterious effect on health, and on this basis industry sometimes argues that therefore no rationale exists for urging that lead should be removed from gasoline. Such an argument misses the point (possibly deliberately). In a region such as San Diego, the burden of lead being contributed to surrounding vegetation is slowly accumulating since it is increasing at a much faster rate than it is disappearing. It is perhaps ironic that the decision to remove lead from gasoline may have been taken by virtue of some short-term considerations relating to the poisoning by lead of certain catalytic devices which some automobile companies were anxious to use on their vehicles.[17] Nevertheless, the evidence on long-term accumulation of this element suggests that its production by automobile exhausts is an important problem, and that the increasing levels which have been tolerated over the past few years could not in any event have been allowed to continue for very much longer. By the time overt health effects had become apparent, it would have taken many years indeed of non-use of lead to reverse such a situation.

Supersonic Transport Aircraft

The problems of air pollution raised by the possibility that future generations of aircraft will fly at altitudes between 65,000 and 70,000 feet have recently been the subject of much public discussion and concern. Provisional estimates were that 500 of these aircraft might be in commercial service by the 1980s and in addition to these there would be the Concordes being made by France and Britain and the Soviet SST's.

A Massachusetts Institute of Technology study group has developed estimates as to the amount of combustion products a fleet of 500 SST's would introduce into the stratosphere.[42] Fears have been expressed that the SST's would put into the stratosphere substantial amounts of water vapour which might reduce or impair the ozone balance and thus reduce the earth's shielding from ultraviolet radiation, or that clouds might be formed which would cause climatic changes on the earth's surface by virtue of their presence at this altitude.[45] The MIT group reported in its conclusions that the operation of aircraft at this altitude might increase water vapour in the atmosphere by 10 per cent globally or by as much as 60 per cent over the North Atlantic, where the SST traffic is expected to be the heaviest. The group concluded that the number of fine particles formed as a result of the operations of these aircraft might be comparable to the amount put into the stratosphere by the volcanic eruption of Mount Agung in Bali (referred to under Particulate Pollution). This load might therefore be expected to have a measurable effect on the earth's temperature. Recent calculations have shown that interference with ozone in the upper atmosphere is unlikely to be a major problem and the effects are apt to be a great deal less than would be caused by particulate pollution.[45]

The magnitude of material produced by an SST is considerable, since during each hour of flight at cruising altitude each SST will produce about 83 tons of water, 70 tons of carbon dioxide, and approximately 4 tons each of carbon monoxide and oxides of nitrogen. There is obviously a great deal more to be learned about the possible long-term effects of the injection of these materials in large quantities at this altitude. No one can pretend that clear-cut answers can be given to this problem, and it is of great interest that the U.S. Senate decided that these and other uncertainties were of sufficient magnitude to halt the further development of the SST in the United States, in the face of the major local employment dislocations and hardships that this decision was bound to cause. However, it is hard to argue with the position that it would be irresponsible to plan for SST aircraft operations of the magnitude projected, unless or until there was a much clearer understanding of the possible implications of the stratospheric pollution they would cause. Some informed consensus will probably be possible on this issue a few years from now, and since the obtaining of the relevant scientific data will probably cost but a tiny fraction of that of developing these aircraft, it seems sensible to urge that this money should be expended first.

SUMMARY

The major public concern about environmental pollution has been centred on two main topics as far as air pollution is concerned; the first is the danger that the population of major cities will be asphyxiated by uncontrolled pollutant emission, and the second is that the long-term effects of air pollutants are so severe that life on the earth will be compromised.

The brief review in this chapter of what appear at this time to be long-term problems, and what can be seen to be "non-problems," will enable the reader to place these matters in some kind of perspective. It must be stated clearly that there are long-term problems and that these cannot safely be neglected. The long-term accumulation of lead is obviously an important subject, and the global build up of particulate pollution or the increasing aerosol concentration at high altitudes, as a result of aircraft operations, are both matters that deserve serious ongoing study.

However, other problems, such as a diminution of oxygen due to fuel burning, and accumulation of CO, do not appear to be major. In this connection it is worth pointing out that the products of automobiles, if lead is eliminated from the exhaust, do not pose long-term problems of much magnitude. The carbon monoxide emitted in large amounts does not appear to be a long-term hazard; the oxides of nitrogen in atmospheric form are not in sufficient quantity to have much influence on the nitrogen cycle of the earth and biosphere; and the hydrocarbons are not abundant enough to have any long-term effects. Only the particulate pollution associated with the operation of automobiles poses a problem in relation to the global environment. For this reason the automobile, although an admitted hazard in a city with high traffic density, static wind conditions, and bright sunlight, is from many points of view an efficient and environmentally acceptable vehicle for long-distance travel.

With the sole exception of particulate lead emission, the automobile, if it can be made safer, does represent a completely acceptable method of transportation across the countryside. Therefore, the wholesale condemnation of the gasoline engine is misplaced. It is urgently necessary to redesign the automobile for city driving or to develop within cities new methods of transportation which do not involve the gasoline engine. But, although this may be freely admitted and urged, both the gasoline engine and the diesel engine will likely remain acceptable and efficient methods of transportation for long distances. The reason is that as far as we know none of the pollutants they emit represent long-term problems; and this after all should be the major consideration.

chapter six

GENERAL PRINCIPLES OF AIR-POLLUTION CONTROL

INTRODUCTION

With the world facing acute problems of uncontrolled population expansion, diminishing natural resources, and pressing socio-economic differences between and within nations, it may well be argued that it is premature to be directing any public attention to the question of air pollution. Apart from the general argument that the existence of other problems is no particular reason not to consider any specific one, there seem to me to be two general reasons why control measures aimed at diminishing air pollution are necessary.

It will have been evident from the preceding chapters that industrial society already pays a considerable cost to alleviate or combat the effects of air pollution in its cities. Although it is difficult to place precise values, in financial terms, on these factors, the first overriding consideration is that in the absence of controls levels of air pollution are certain to become worse and eventually difficult to manage. For example, the tonnage of sulphur dioxide released in Montreal was estimated in 1961 to be approximately 550 tons per day. By 1966 this estimate had risen to 775 tons per day. Most recently the figure is computed from a detailed study of air sampling in the region to be closer to 1870 tons per day in winter, and approximately 1200 tons per day as an annual average.[35] The growth of a city, which has to heat itself with fuel oil, is bound to be accompanied by increasing levels of atmospheric sulphur dioxide, unless the sulphur content of fuel is controlled.

The same kind of projection can be made concerning oxides of nitrogen. One of the most recent emission estimates, based on present legislative standards for the United States, predicts an increase in oxide of nitrogen emission from cars from just over 7 million tons per year in 1970 to about 12 million tons per year in 1980, and to 17 million tons per year in 1990.[63] Such projections indicate the inevitability of control in the future if the urban atmosphere is not to become intolerable.

Furthermore, levels of oxidant air pollution in the Los Angeles region are very close to those which produce overt symptoms and some measurable changes of pulmonary function in the population, [13,16] and in the absence of future controls these concentrations will almost inevitably rise. The problem of accumulation of lead in vegetation and in the environment points to another much longer-term problem, which certainly calls for careful study and consideration if ambient levels are not to reach a concentration too high for the future health of the population.

The second reason why consideration of air-pollution control is pressing is that the extension of industrial conditions of the western world to countries and cities which at the moment do not have much industry or many automobiles will inevitably cause exactly the same problems in those regions as are now being encountered in the West. As it is by far the most affluent society that the world has yet known, the industrial West has to take the lead in research and development to enable those countries developing somewhat later to avoid the difficulties which are now being encountered in Los Angeles and New Jersey.

As soon as it is admitted that some controls of certain pollutants are necessary, a number of other very important considerations immediately present themselves. These concern the technical ways in which air pollution may be controlled; the collaboration between government, industry, and the public, in discussing which control shall be used, and in what order of priority; and, more broadly, the social considerations which have to underlie such decision-making. Such decisions will vary depending on the kind of society in which they are being made, and it is not surprising that different approaches have been taken by different countries to the overall problem of pollution control.

In this chapter, a very brief summary of some of these concerns is attempted, but the reader must recognize that the whole field of pollution control is moving so fast that whatever is written today is likely to be made out-of-date by tomorrow's newspapers. During the past two years, however, some general trends which are deserving of study have emerged in different countries, and some experiences in relation to legislative enforcement have been acquired, as a result of which it is possible to see some of the strengths and weaknesses inherent in different kinds of programs. Canada has been faced with some unusual problems in this regard, since traditionally there has been a major jurisdictional separation of function among cities, provinces, and the federal government.

At a more general level, the attempts by society and governments to deal with the total problems of the environment, as these are now understood, have brought to light serious deficiencies in decision-making procedures. Methods of governing and decision-making appropriate fifty years ago do not now seem capable of tackling many of the issues of contemporary society, and the dilemmas raised by air pollution provide an example of the acute strains and stresses to which the governmental decision-making process has been subjected in the past ten years. The problem can be tackled by breaking it down into manageable parts, and by forming some provisional views on ways in which further environmental degradation can be avoided and a better air and water environment handed on to the next generation than was inherited by this one.

In this continuing process, one of the most important

features is the education of future generations in a mode of thought different from that in which their parents were brought up; this can now be seen as one of the main tasks of the present to protect the future. It is to be expected that the brief era of polemic and gloomy prognostication through which we have passed will be replaced by one in which the less exciting, but more durable, tasks of translating intent into actuality call upon the concentrated effort of scientists, lawyers, and politicians. There is no doubt that most responsible individuals, whether they work in industry or in universities or in government, understand the necessity for this kind of thinking. The statement made by the president of the Gulf Oil Corporation in the *Wall Street Journal* of 22 March 1971 is representative of this new dimension in corporate thinking. He said: "Maximum financial gain, the historical No. 1 objective of American business, must in these times move into second place—whenever it conflicts with the wellbeing of society."

Many intervening steps and decisions must be taken before this progressive attitude results in environmental improvement, but few people now doubt that the will to take these steps exists throughout a sufficiently broad sector of society, and in adequate strength, to ensure that this shall be realized.

METHODS OF CONTROL

Control of Specific Emissions

There are a great number of regulations for many different countries controlling the emission standards for specific pollutants. Most of the European countries, including Czechoslovakia, and the USSR, together with Japan and the United States, have many emission standards for specific pollutants discharged from particular industrial processes.[5] These cover solid particulate pollutants of stated size, metallic particles, and organic chemicals in wide variety. In some areas the dust concentration permitted in effluent gases is specified, and in many countries the height of a chimney stack is outlined by regulations. The alkali inspectorate of Great Britain, for example, specifies a certain basic chimney height for any given efflux of gas (in feet per second) in relation to the evolution of sulphuric acid.[5] Similar regulations exist in Belgium, Czechoslovakia, and other countries. Many existing international data on emission standards are summarized in the third volume of the textbook edited by Dr. Arthur Stern.[5]

In general, these standards for such substances as hydrogen sulphide and oxides of nitrogen have been arrived at as being the maximum permissible level which, it is calculated, will not produce harmful effects on vegetation, or which will not be a major nuisance to those living adjacent to the plant. There is however considerable variation in the standards which are considered either "basic" or "permissible." One might remark that in the USSR the basic standard for hydrogen sulphide, for example, is 0.005 ppm and the permissible standard is also 0.005 ppm. In Pennsylvania the basic standard is 0.005 ppm, as in the USSR, but the permissible standard is a concentration of 0.1 ppm.

The variation of permissible emission standards for sulphur dioxide also shows very considerable diversity, by at least a factor of ten. It may be noted here that, although many of the industrial processes referred to in this kind of legislation take place in Canada, there is a conspicuous lack (indeed a virtual absence) of basic or permissible emission standards applicable to the whole of Canada. The reasons for this are noted in the section below on Air-Quality Criteria. There are, however, some emission standards for the province of Ontario covering such substances as ammonia, beryllium, carbon disulphide, carbon monoxide, chlorine, and fluorides amongst others.

In Great Britain, these emission standards have been developed on the basis of legislation introduced in 1863 on the recommendation of a Royal Commission to deal with the hydrochloric acid evolved during the manufacture of alkali.[50] The "Alkali Act," as it came to be called, has provided a basis for a whole series of regulations and a total of fifty-six different industrial processes are now on a schedule in which not only is the emission of substances controlled, but the inspectorate has wide powers of inspection, testing, and enforcement. There is little doubt that in the United Kingdom this body of legislation, together with the inspectors who enforce it, has provided a reasonable control of processes which result in the evolution of potentially harmful substances.

Although emission standards may be specified by a country, it it is hard to be sure in some instances what steps are taken to enforce them, or to make sure that the necessary conditions are being observed. The USSR has some stringent standards in relation to industrial effluents, but it seems likely, from some of the published data on air pollution, that in many places these conditions must have been waived in particular instances, or the inspection process must be somewhat loosely enforced.

During the last few years there have been major advances in the technology required to limit industrial effluents. New processes to deal with major particle pollution have been developed, new absorption techniques have been brought into operation to capture such substances as acetaldehyde and formaldehyde, and the engineering needed to exert control of sulphur dioxide and oxides of nitrogen has been the subject of much research. It is important to note that the general public's demand that gross sources of air pollutants be controlled has in its turn led to the proliferation of research and development in the field of pollution control. The results of this are already apparent in better methods available to industry to deal with many of its more severe problems. A full discussion of present control methods in relation to specific industries is beyond the scope of this volume, but excellent summaries of existing technology will be found elsewhere.[5]

Control of Emissions from Cars

It has been recognized for some years that automobiles, as used in an urban area, present special problems of pollution control. Since the problem of photochemical-oxidant smog secondary to automobile-exhaust emission was first recognized and studied in detail in Los Angeles, and it is in this region that it has reached in general the highest levels, the state of California was an early pioneer in setting standards for vehicle-exhaust emissions. An uncontrolled car may be expected to put out in its exhaust approximately 900 ppm of hydrocarbons, 3.5 per cent carbon monoxide, and 1,500 ppm of oxides of nitrogen. The California standards for 1966 specified that these emissions had to be reduced to 275 ppm of hydrocarbons, 1.5 per cent of CO, and 350 ppm of oxides of nitrogen.

The first source of car emission to be controlled was the liberation of crankcase gases into the atmosphere. Gases escape from engine cylinders around pistons and hence reach the crankcase. If there is no control, these gases account for about a quarter of the car's emission of hydrocarbons. By use of a simple device, the escaping gases can be cycled back into the engine (a system known as "positive crankcase ventilation") and the emission of hydrocarbons very much reduced thereby. A redesign of carburetors to limit the evaporation of gasoline directly from them also led to a considerable reduction in emissions, and a redesign of carburation was responsible for achieving a diminution in carbon monoxide levels. The 1970 California standards call for a further reduction in hydrocarbon (to 180 ppm) and carbon monoxide emission (down to 1 per cent) and the 1975 projected regulations call in addition for a drop in oxide of nitrogen emission.

The discharge of lead can only be effectively controlled by using gasoline which does not contain this particular material. A recent review of the general problems of automobile pollution indicated that a number of reactor devices were being brought forward to treat automobile exhaust, and at least one of these appeared to make it possible for the automobile industry to meet the proposed 1975 U.S. Department of Health, Education, and Welfare standards for automobiles, and to have reduced the output of nitrogen oxide to only one-third of the 1974 California requirements.[17] The feasibility of this device in ordinary automobile usage has yet to be proven.

There is obviously much further work to be done, but it has become clear that the survival of the gasoline engine essentially depends on the ability of research engineers to produce a vehicle with very low emission standards for the three main pollutants produced by motor cars. Since noticeable progress has been made in this direction during a short period (in relative terms) within which considerable attention and money have been devoted to this problem, it may be that a suitable vehicle for the urban environment will be produced in the next few years. More radical solutions to the problem, involving entirely new energy devices such as electric power or engines of completely new design, have not yet been shown to be feasible as competitors for the gasoline engine, but in a few years such developments may well make the existing automobile engine seem as obsolete as are the commercial aircraft of only ten years ago.

Actual control of automobile emissions is difficult to put into practice. In a country such as Canada, not only are there complex jurisdictional problems, which will be referred to later, but it is not easy to be sure that devices placed on automobiles at the time of their manufacture either stay on them or remain in efficient working condition. An elaborate system of testing stations capable of measuring the effluents from a car's exhaust has been proposed, but would be costly. In many states of the United States, cars have to be inspected annually for mechanical performance, and this would give an opportunity for the emissions from an automobile to be checked.

There are as yet no definitive emission regulations for diesel vehicles, and it is a matter of common observation that many of these emit a considerable amount of smoke; the manufacturers of diesel engines state that this should never occur if an engine is properly adjusted and once it has been warmed up. In some countries there are regulations which permit a diesel vehicle or an automobile to be stopped by the roadside if its emissions are judged a nuisance, and in recent years there has been less reluctance to enforce this kind of regulation. Few countries in Europe have yet adopted emission standards for automobiles, as they have judged that there is no direct evidence that automobile emissions constitute a hazard to health under present conditions in their countries. It seems likely that this outlook will be modified as soon as technological advances have permitted relatively cheap modifications of the gasoline engine which will bring its emission standards down to an acceptable level.

The Province of Ontario in Canada has introduced regulations concerning air contaminants from motor vehicles and specified the level of hydrocarbon emission for engines of different size. [6,71] For a new motor vehicle operated in Ontario, where the engine displacement is at least 50 but not more than 100 cubic inches, the emissions shall not exceed 410 ppm by volume of hydrocarbons or 2.3 per cent by volume of carbon monoxide. Where the engine displacement is more than 100 but not more than 140 cubic inches, the emissions shall not exceed 350 ppm by volume of hydrocarbons or 2.0 per cent by volume of carbon monoxide. Later regulations deal with smoke discharge from diesel vehicles and establish procedures for testing whether or not these are being exceeded. These ordinances do not specify levels for oxide of nitrogen emission and, as most recently promulgated, refer only to new cars manufactured after the beginning of the 1970 model year and sold, offered, or exposed for sale to or used by the ultimate purchaser who is a resident of Ontario. These regulations have not yet been extended to cover other provinces in Canada or the nation as a whole.

Emissions from Domestic Sources

Many cities specify certain conditions for the operation of incinerators in apartment blocks, and most of these regulations are designed to control the black smoke often given off and the fly-ash nuisance to which they may give rise if they are improperly operated. It may be pointed out, however, that the total pollution from domestic incineration is extremely small. An estimate for the total United States, for example, concluded that refuse disposal as a whole, including municipal, accounted for only 3.5 per cent of total pollutant tonnage; a calculation for the city of Montreal for 1966 estimated that municipal incineration, which then amounted to 100 tons a day, was responsible for the emission of only 5 tons of solids (out of a total of 206 tons per day for the city as a whole), and negligible quantities of sulphur dioxide and nitrogen oxides. Domestic incineration was estimated to be responsible for a further 3 tons of solids emitted per day (see table 3).

These are small totals in relation to the urban problem, yet by being conspicuous the plume of smoke from an apartment block may be a common source of complaint to a city health department. It must also be recognized that consistent violation of regulations may constitute a serious nuisance to an apartment dweller whose window overlooks at close range such a chimney. As far as domestic fuel is concerned, it has proved much easier to control the fuel than to control the emissions, and the limitation of the pollution produced by domestic heating has essentially been by fuel control.

As a result of the December 1952 smog disaster and the Beaver Report which followed it in 1954, a Clean Air Act was introduced in the United Kingdom in 1956. This is operated by local authorities and is concerned with fuel from domestic and industrial sources. It became an offence to emit dark smoke (which was defined as smoke darker than a certain standard known as Ringelmann No. 2) for longer periods than permitted by regulations made by the Minister under the Act. The local authority was empowered by an Order, confirmed by the Minister, to declare a district to be a "smoke control area" in which only approved appliances or approved fuels could be used. What this did in fact was to make it illegal to burn coal on an open domestic hearth in the City of London, and any local authority could make the same regulations in respect to other areas of the country to control the domestic emission of coal smoke. The only alternative fuel readily available at that time was coke, and solid fuel of that kind came to be used very generally as a substitute for coal. This has had the effect of greatly reducing the smoke level in London (see chapter 7) although, since coke still contains considerable amount of sulphur, the sulphur dioxide emission rate has been changed much less (see figure 8). In retrospect, this must be regarded as having been a very successful kind of legislation to deal with an immediate situation.

Cities which depend on oil for heating have moved to control the sulphur content of fuel which may be used. In Montreal, the sulphur content of different grades of fuel oil was controlled by city bylaw,[25] with a phased reduction in permissible sulphur content of fuel oil: light domestic oil was not allowed to exceed 0.6 per cent sulphur in 1970, will not be permitted to exceed 0.5 per cent in 1971 and 0.4 per cent in 1972. Heavy oil used in larger heating plants had a maximum allowable sulphur content of 2.5 per cent in 1970, and will not be permitted to exceed 1.5 per cent in 1972. The Ontario regulations made under the air-pollution control act specify the oil by numerical grades, and also call for a reduction in heavy oil from 2 per cent sulphur content in 1971 to a 1.5 per cent in 1973.[71] The lightest grade of oil is permitted a 0.5 per cent sulphur content until 1973.

Other cities, particularly in the United States, have similarly controlled the sulphur content of fuel oil, and the sulphur content of bituminous coal being used for power generation is now also carefully supervised and specified. In a city with a cold winter, such as Montreal, the contribution of sulphur dioxide from fuel oil used for heating becomes a major pollution source during the winter.[35] The control of the sulphur content of fuel oil has the advantage that it is easy to enforce since the manufacturer or distributor is required to submit samples for analysis, and it is not difficult to determine the sulphur content. Infringement of the regulation carries severe penalties, and the city is empowered to analyse any sample of fuel oil being sold within the city for consumption. This kind of legislation also has the advantage that it can be made selective. Fuel oil of a higher sulphur content for domestic purposes may be considered permissible in country areas, since the concentration is unlikely to do much damage to vegetation. Thus the legislation can be made flexible, and the restrictive clauses confined to specified urban regions where the reduction of sulphur content is of much more importance.

AIR-QUALITY CRITERIA

A somewhat different approach from those described in the sections above has been taken in the United States. The Air Quality Act of 1967 required of the Secretary of Health, Education, and Welfare that he should "from time to time, but as soon as practicable, develop and issue to the States such criteria of air quality as in his judgment may be requisite for the protection of the public Health and Welfare. Such criteria shall . . . reflect the latest scientific knowledge useful in indicating the kind and extent of all identifiable effects on health and welfare which may be expected from the presence of an air pollution agent. . . ."[14] This directive has led to the publication of a number of monographs [9, 10, 11, 12, 13, 14] dealing in considerable detail with specific pollutants and providing the individual states with basic guidelines for decision-making concerning the advisability or otherwise of introducing air-pollution legislation of one kind or another. The monographs have provided most valuable summaries of existing knowledge, and although some of the conclusions they have contained have been somewhat controversial, as they were bound to be, they have provided a basis for some legislative action.

Concern for environmental problems in the United States has more recently led to the establishment of an "Environmental Council" with wide responsibilities of an advisory nature, and the bringing together of a number of scattered agencies into one major bureau known as the Environmental Protection Agency. This is responsible for dissemination of abstracts of world literature on air pollution, as well as the publication of the monographs on air-quality criteria being developed by the U.S. Department of Health, Education, and Welfare.

There has been major debate in the United States in the public sector on how to establish adequate criteria for air quality. Though some of this is repetitive, much of it is worth reading since in few other countries has so much public evidence been laid before senior legislators and so much discussion occurred on how to go about assessing available evidence and arriving at suitable standards.[46,47]

The Air Quality Act of November 1967 also prohibited state and local governments from regulating emissions from new motor vehicles, and made this a federal responsibility, although provision was made for this prohibition to be waived in any state which had adopted vehicle emission standards before 30 March 1966 (essentially this meant California only). A strong federal program of air sampling was also established, and the federal agency was given power to make measurements of air quality and to publish these, without being subject to the jurisdiction of the state.

A distinction must be made between air-quality criteria and air-quality standards. The criteria are intended to be the basis for judgement as to what air quality should be aimed at, whereas an air-quality standard should state rather precisely the time/concentration relationship permissible to that community for any given pollutant. Thus if it is stated that, on the available evidence, 0.02 ppm of SO_2 as an annual average can be considered to be without any demonstrable adverse effects on health, an air-quality standard might be established at 0.02 ppm as an annual level, 0.1 as a daily 24-hour maximum, and 0.3 as the highest peak concentration which would be permitted in a given city.

In Ontario, the annual geometric mean of soiling index for industrial commercial regions has been established at 1.1 COH units/100 feet of air, and a level of 0.45 COH units/100 feet of air for residential and rural areas. As noted earlier, there are many regulations in different countries concerning the height of chimney stacks for industrial processes discharging sulphur dioxide or particles, and these have generally been computed on the basis of some minimal air-quality standard downwind of the plant.

Although some other countries have been reluctant to follow the United States lead in trying to establish air-quality criteria, there is little harm in their being discussed and in making attempts to reach some definitive statements as to what level of pollutant should be considered permissible. The difficulty of establishing such levels may be admitted, but it seems to me better to attempt to define acceptable levels, and work towards these, and be prepared to modify them, than to dismiss them as too idealistic and too contentious for any legislative action. Unfortunately many of the kinds of study, even of the short-term effects of pollutants, which are required to establish these values have not yet been made, and the scientific literature contains very few laboratory studies carried out with an adequate number of subjects and under sufficiently realistic conditions to permit conclusions to be drawn.

Yet it may be pointed out that pollution control aimed only at emission control from some specific industries and from domestic fuel leaves unanswered the general problem of automobile pollution, or the additive effect in a major urban area of multiple pollution sources. It may be that piecemeal legislation will be effective in the long term at dealing with these problems, but it seems more likely that, as more knowledge is gained, a wider legislative field of action would be considered desirable.

I have little doubt that the concentration of attention in the United States on trying to develop air-quality criteria for the common pollutants has been an extremely valuable device for focusing attention on areas of ignorance, and in trying to develop the best consensus possible to guide the different communities in the legislation they may adopt for air-pollution control. The very considerable budget which was placed behind the development of these criteria has led to the growth of measurements of air pollution in different American cities— of tremendous value to the international community as well as locally (see table 4 for example).

Proposed Air Quality Standards for Canada have recently been tabled in the House of Commons and are given in the Appendix.

LEGISLATIVE ACTION AND ENFORCEMENT

In some countries such as Holland and Sweden, the initiative and responsibility for enforcement have been taken primarily by the central government of the country, though certain aspects of monitoring and control may be delegated. In Great Britain, the responsibility for inspection and enforcement of the Alkali Acts controlling industrial effluents rests with an inspectorate division which is answerable to the national government. The local authorities at the county or city level have powers vested in them through the Clean Air Act relating to domestic fuel. So far there is in Britain no central parliamentary legislation dealing with control and inspection of automobiles.

In the United States, the interrelationship between the federal government and the individual states is not only variable and complex but is undergoing more or less continuous revision. As noted in an earlier section, the federal government has accepted the responsibility for automobiles in all states except California. It also has the jurisdiction for major air monitoring programs covering different cities, and it has passed considerable enabling legislation giving individual states powers to control air pollution in general. Where the new Environmental Council will fit into this pattern is not yet clear, but some central overseeing authority with considerable power to intervene may be necessary if state or city governments are not to be held to ransom by individual industries.[41]

Canada has been faced with very complex jurisdictional problems which cannot be discussed in great detail in this volume. International problems, mainly related to the air pollution "drift" between Detroit in the United States and Windsor in Canada, have been recognized as within the jurisdiction of the federal government. It has been clear that municipalities such as Toronto and Montreal have the power, within existing bylaws, to limit the sulphur content of fuel or the operation of incinerators within their boundaries; and both have used this. It has also been apparent that individual provinces such as Ontario or Manitoba may control effluents from industries situated within those provinces, but the federal government has not brought this legislation into any coherent pattern.

The St. Lawrence River, as an international waterway, comes under the jurisdiction of the Department of Transport, which can, at least in theory, prosecute anyone on it in a rowboat if he is not carrying DOT-approved life jackets. This same Department of Transport, however, has not been considered to have any authority over motor cars on Canadian roads, since these are not "international." It has not been at all clear in whose jurisdiction legislation about motor vehicles really lies, or who is prepared to implement such legislation if it were to be introduced.[31]

The new Canadian Clean Air Act[1] steers a middle course between the British and American approaches to air-pollution control—following a precedent of much previous Canadian

legislation.[79] Attempting to combine the principle of control
via available technology, and the philosophy which seeks to
deal with air pollution essentially as a management problem of
air resources, Bill C-224 endeavours to promote some sort of
uniformity across Canada. The Bill tries to complement varying kinds of provincial legislation and to bring all these finally
into line one with another, to avoid the possibility that some
provinces might permit levels of pollution much greater than
others. The Bill seeks to permit the Minister to develop uni-
form national air-quality objectives, defining these in general
in four ranges for each pollutant.[1] (See Appendix.) It also en-
ables the federal government to prescribe national emission
standards and permits the Minister to make regulations under
this paragraph. It is likely that initially only extremely hazardous pollutants will be considered under this section, but it is
not easy to predict how the powers so conferred for the first
time will in fact be used. The Bill also sets up for the first time
a national air-surveillance network, and permits the federal
government to enter into agreements with individual provinces
to conduct joint cooperative control programs; on the basis of
such accords the federal government is empowered to take
direct action in certain emergency situations (which seem very
unlikely to arise in actual fact).

There seems little doubt that in most countries there is a
considerable sphere of operation in air-pollution control for
the local government, presumably coordinated with a city or a
region, a provincial government, and a central government.*
In small countries all of these functions can be combined, but
in larger geographic units it is obviously sensible that the
powers should be divided in a practical way. Enforcement of
municipal bylaws is clearly the task of a city, and in Canada
this is undertaken, for example, in Toronto and Montreal.
Control of a major effluent problem in an industry situated
outside the city limits, but providing a considerable burden of
pollution over a city region, has to be vested primarily in the
province, but such provincial legislation has to be in line with
statutes elsewhere in the country or internationally.

*I understand that the municipal government of Tokyo has insisted
on stricter standards for some air pollutants than those prescribed by
the national government.

The ubiquitous motor car cannot be controlled by a city (though it can be excluded from certain streets in the city) and there are major problems of inspection and enforcement which will have to be operated in most areas if legislation is to have very much effect. In many countries, as noted later, air pollution is raising international problems, and these would have to be dealt with on behalf of a country at a federal or central level.

In this complex situation, it seems unlikely that any single legislative pattern will come to be adopted by all countries. To some members of the public, keen to see that careful consideration is given to overdue control of some serious sources of pollution, the discussions of jurisdictional responsibility may seem interminable. Everyone is familiar with the fact that a division of responsibility very often means that no single person accepts any obligation for anything, and it is very easy for different levels of government to shelter behind a confused jurisdictional situation as a justification for general inaction. There may even be isolated industries anxious to see that the situation remains as confused as possible for as long as possible, since they rightly sense that this creates a circumstance in which any effective supervision or control of a particular function is very unlikely to become operational. This predicament is exemplified by much of the footwork and jockeying which has been well documented by Ralph Nader as having occurred in relation to specific pollution problems in the United States.[41] It seems quite clear that one of the ways to overcome this kind of problem is to insist on certain stages of decision-making to be public. This point is emphasized in the next section.

DECISIONS ON PRIORITIES

The immense ferment of public discussion on the environment and its deterioration led very quickly to detailed consideration of how decisions should be made in a modern society on the priorities of national concern, and how government decision-making should properly be influenced. With a perception that God seems to give to the young and deny to the middle-aged, the student population realized very quickly that this kind of challenge would expose structural weaknesses in a technologically oriented society in which collusion between business interests and government had traditionally formed the basis for a great deal of policy-making. More significant even than this was the fact that the traditional methods of exerting public pressure on decision-making by government —mainly by expecting that the public would show its approval or disapproval by re-electing the government after a span of office of five years—could be seen to be far too slow and diffuse a process in relation to many of these decisions. The ability of the news media to generate public concern is formidable, and, although a great deal of misleading information gets disseminated in the process, the politician has in general been made aware that there are legitimate grounds for public concern on the deteriorating condition of the air of the city in which he lives, or the water which is available to him for recreational purposes.

An additional complicating factor has been the traditional dislike of civil servants to have the public intervening to any major extent in what are often regarded as matters which could concern only government, civil servants, and industry. It has proved difficult for governments to broaden the base of decision-making, and even now there is reluctance to allow the public to be party to data relating to air pollution (which in almost every instance have been collected at the taxpayers' expense).

In Sweden there does seem to have been a serious attempt to involve the public as a whole in the decision-making processes, and to place adequate executive powers in the hands of a concessions board, which has authority to set limits on how much pollutant may be put into the environment.[43] This board consists of a chairman trained in the law, a technical specialist in the area of pollution, and two persons without any particular affiliation. In addition to this organization, there is a consulting board for environmental problems whose twenty-four members include ten scientists plus delegates from labour, industry, finance, and the press. This represents an unprecedented involvement of members of the public in governmental councils, and no other countries have had the courage to follow such an example.

The hearings of a Senate committee in the United States [46, 47] served as one kind of outlet for legitimate expressions of public and scientific opinion on specific pollution issues, and it may be that the Environmental Council in the United States will perform some of the functions of its Swedish equivalent. In most instances, however, politicians have been unable to see beyond the traditional mechanisms they inherited for decision-making on these issues, and the consequence has been that many members of the public are left with an uneasy and justifiable suspicion that if matters are to be discussed only between industry and government it is very far from assured that the public interest will be properly protected.

Professor J. K. Galbraith in his book "The New Industrial State" points out that economists are bound in the main to be the natural allies of the industrial system, and places the responsibility for the re-orientation of society on educated members of the community. He states: "In fact, no intellectual, no artist, no educator, no scientist, can allow himself the convenience of doubting his responsibility. For the goals that are now important there are no other saviours. In a scientifically exacting world scientists must assume responsibility for the consequences of science and technology. . . . The individual member of the educational and scientific state may wish to avoid responsibility, but he cannot justify it by the claim of higher commitment."

Whereas such sentiments are very proper, if unexpected, from an economist, it has to be pointed out that in very few areas are there proper channels for the expression of a sense of responsibility except by publicizing matters through newspapers, or apparently being inevitably committed to an attitude of hostility towards government. For this, government has only itself to blame, since it has structured few other channels of consultation and discussion through which such opinions can be freely expressed.

The question therefore of who should decide the priorities remains unanswered. In the last analysis it is bound to be the government, but the extent to which a government may or may not be sensitive to the public wishes, and may be able to be independent of powerful industrial interests, remains quite uncertain. Since however the cost of controlling pollution adds directly to the cost of living—or put another way, the cost of controlling pollution detracts inevitably from average annual income—it seems proper to insist that the public should have some say in its priorities.

It has been of particular interest that the accounting profession has recently been concerned with its inability to place even any approximate estimate of amenities into a proper accounting framework.[75] It will be of great interest to see whether any attempts to catalogue amenities or loss of them can be put on some reasonable basis. How is one to quantify a degree of air pollution sufficient to cut visibility in Manhattan to a mile and a half, and make hazardous the operation of La Guardia airfield? How is one to cost the eye irritation and smarting that occurs in Los Angeles? It may be possible, as Lave and Seskin have attempted,[58] to make some approximate estimate of secondary health costs in relation to air pollution and, as noted earlier, some assessments can be made of material damage. But no one would pretend that these can properly be considered to comprise the whole estimate of amenity loss in relation to air pollution. Possibly social scientists or management consultants in the future may be able to arrive at some agreed system of evaluation which, after trial and error, will give some idea of the loss of amenity which occurs in the presence of severe air pollution.

ACHIEVEMENTS AND FUTURE NEEDS

The achievement of the city of Pittsburgh in diminishing to a great extent its air pollution during the years just before and after the war will long remain a model of what can be done. More recently, however, there have been other striking examples of improvement in the environment as a result of legislation. In figure 8 are shown levels of sulphur dioxide and smoke in London from 1958 to 1968. It will be remembered that the Clear Air Act was introduced in 1956. The emission

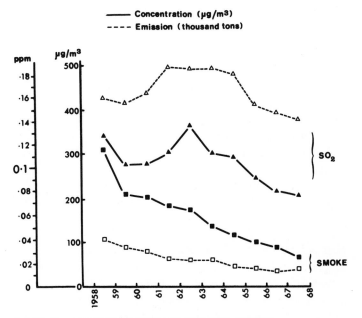

FIGURE 8. CHANGES IN AVERAGE SULPHUR DIOXIDE
SMOKE LEVELS IN LONDON SINCE 1958

The Clean Air Act, which was introduced in 1956, effectively prohibited
the open-burning of coal in the greater London area. It is of interest
that there has been a great reduction in smoke concentration since
1958, and although SO_2 emissions have not altered greatly, the average concentrations recorded have fallen very significantly. This is
attributed to the better "ventilation" of the city as a consequence of
the major reduction in particulate pollution. The second ordinate scale
refers both to $\mu g/m^3$ and to thousands of tons in relation to emissions.

of smoke has fallen to less than half its previous level, and the
resulting concentration (in $\mu g/m^3$) has fallen from 300 to
approximately 70. It will be remembered (see figure 6) that
during the smog episode of 1952 the smoke levels were above
1000 $\mu g/m^3$ for a period of four and one-half days. The sulphur dioxide *emissions* have fallen slightly, but it will be noted
that the sulphur dioxide *concentrations* have fallen proportionately more, since in 1958 levels were running at 0.13 ppm and
by 1968 had dropped to about 0.08 ppm. The greater drop in

SO_2 average concentration than in the rate of emission is believed to be explained by the fact that the much lower smoke concentration has meant not only more hours of sunlight in the city, but much better convection ventilation, with a consequently more rapid dispersal of sulphur dioxide from ground level.

These improvements in London are already being reflected in changes in morbidity. Not only have there been no episodes of comparably severe pollution, but patients with chronic bronchitis studied over the winter of 1964–65 showed less response to changes in pollution on a day-to-day basis than did those who were studied five years before. A long-term study of bronchitis among middle-aged men working in West London between 1961 and 1965 is said to have demonstrated a decline in the amount of sputum they expectorated during the period when there was a fall of concentration of smoke in central London. As noted earlier, a general reduction in level of pollution in terms of average monthly or annual rates is bound to be reflected by a fall in peak concentration (see figure 1). For this reason, steps taken to lower the overall level of pollution are almost certain to be effective in reducing the average pollution level.

Similar improvements have been recorded for Manhattan as a result of legislation limiting the sulphur content of fuel oil and paying attention to the sulphur content of coal being used to generate electricity. In figure 9 are shown data for Manhattan for the years 1965–1969. From this information it is clear that whereas a level of 0.5 ppm of SO_2 was exceeded for 300 hours in 1965, this level was exceeded for only about one-tenth that time (or 30 hours) in 1969. This represents a major improvement in the air of downtown Manhattan entirely attributable to the controls that have been introduced.

Serious though the problems of automobile pollution have been in Los Angeles, the state of affairs that would now exist if controls had not been exerted on automobile emission would have been much more grave. Although the general level of pollutants in the Los Angeles area as a result of automobile operation has not yet begun to fall, the increase in automobile traffic which has occurred over the past ten years, in the absence of control legislation, could have led to levels of air pollution which would now be catastrophic. These developments clearly indicate that the possibility of control with existing legislation is not remote, but fairly easily obtainable. They do not lend any support to the view that no action can be taken within the existing framework of decision-making to improve the situation. Admitting that there are long-term problems, particularly

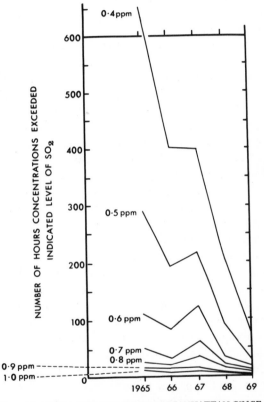

FIGURE 9.   SULPHUR DIOXIDE LEVELS IN MANHATTAN SINCE 1965

Data from reference 37. As a result of legislation affecting the sulphur content of coal used for electricity generation, and also control of the sulphur content of fuel oil, there has been a striking reduction in the number of hours in which certain concentrations of SO_2 were exceeded in New York in 1969 compared with 1965. In general, there has been a tenfold reduction in time for any specific concentration level of SO_2 over this period.

of energy resources, which are very far from being solved, there seems little doubt that the immediate short-term problems can easily be handled once there is a real intent to do so.

It is possible from this vantage point to look forward and to specify some needs for the future which are immediately apparent. These go somewhat beyond the immediate problem of air pollution, but they encompass it. An incomplete listing would include the following items.

1. There is an urgent need for complete revision of our educational programs, starting in elementary school, continuing through high school and university, to ensure that the next generation is brought up with an understanding of living processes and of the biosphere as a whole, so that the proper questions are asked in advance of the impact of any projected change on the environment. The present generation of adults was not brought up with this philosophy, and the future can o. be well served if the present children are brought up and educated with a strong understanding of the meaning of the interrelationship between different parts of the living world, the broad essentials of the support of the biosphere, and the ways in which degradation of the environment can be avoided.

Adult education programs are necessary at the moment because of a very serious deficiency in this aspect of education over the past thirty years. As soon as first-class educational programs are being given at school and university level, the need for adult education will in large part have disappeared. In parts of the English-speaking world biology has traditionally not been regarded as a "hard science subject." We have been brought abruptly up against the hard fact that it is unquestionably a subject basic to human survival.

2. Continuing attention has to be given to how the public can be involved in proper priority decision-making. The first essential in understanding this process must be a changed outlook in regard to giving members of the public at large the scientific information relating to all aspects of pollution. Many data on pollution are freely obtainable, and it is greatly to the credit of the United States that it has made available far more data relating to its own pollution than has any other country in the world.

Elsewhere, there is a reluctance to allow publication or to give publicity to any collected information, and it is therefore exceedingly difficult to come by. A necessary part of this process of decision-making is that data on effluents, which the public can see coming out of different industrial plants, must be freely available to an inquirer; government at the municipal and city level should have the power to require such industries to declare what effluent load they are discharging on a 24-hour average basis. There seems to be no good reason why the public should have to go to great expense to find out this information which is of great interest to it. Such data might well be collected by an environmental council and placed intermittently in the public sector.

3. Another major future need is to broaden the base of decision-making when major policy questions are confronted by government and industry. Many inputs must be taken into account when the particular problems of an industry, or of an automobile, for example, are being considered, and there is no way that proper decisions which bear on public need can safely be left to government and industry alone. My view is that it is the responsibility of government to structure new mechanisms to enable the public to take part in decisions which vitally concern its welfare and the future environment of its children. Although I can fully understand the reluctance of government to broaden the basis of decision-making in this way, I can see no other alternative as an acceptable way of proceeding. It may of course be argued that decisions relating to the national budget are not taken in this way, but by the traditional method of expecting the public to vote out of office a government whose resolutions in the main it has disliked.

As noted earlier, the kinds of decision-making required to determine at what economic cost modifications of the automobile ought to be insisted upon, or whether present levels of pollution in any particular city justify major expenditures, are a somewhat different type. They involve a series of value judgements which accountants and economists cannot yet make with any precision, and as such they should be subject to a much broader basis of discussion and decision-making than government thus far has been prepared to allow.

4. It should be evident to the reader that there is a continuing need for research. Members of the public who become impatient that so little is apparently being done in their area to cope with what they see to be major problems often state that there is no point in spending more money on research. One can sympathize with this attitude, but in the whole discussion of pollution there is really no substitute for factual information. The imperfection of current knowledge relating to the effects of present pollutants, and the importance of studying materials which may well become important in the future, all necessitate continued expenditure on this item of the public-health field.

Industry may be expected to invest money into research on the technology required to control pollution better, and it has a reasonably good record of doing this. It cannot be asked to finance studies of morbidity or mortality in relation to pollution; this must be the responsibility of other agencies. The very large amount of money which the United States has disbursed for research on all aspects of air pollution is a major investment for the benefit of the future of the industrialized West, and many important observations have been derived directly as a result of this expenditure. The per capita outlay in the United States on this type of research is vastly greater than in any other western country and this fact alone is greatly to its credit.

5. Finally, one can see the necessity in air-pollution discussions of paying attention to the point made by the Swedish government in its appeal to the United Nations to have a conference in 1972 on environmental-pollution problems. Perhaps it is not surprising that Sweden should have initiated this, since it has been in the vanguard in many aspects of pollution research and legislation. Yet the purpose of calling the conference together was to consider ways and means whereby international cooperation in pollution control could be brought about. It is quite apparent that present commercial structures will inevitably favour an industry which can produce its product without any control of attendant pollution. Thus, for example, if in Sweden the pulp and paper industry is required to exercise strict control of its water- and air-pollution effluent,

whereas in another country such as Canada the industry has no such requirement, it is apparent that paper will almost certainly be produced more cheaply by the industry not subjected to strict pollution surveillance.

Another example whereby a "cut rate" kind of operation may favour pollution has occurred when oil tankers have been operated without a strict regard for safety features in order to cut costs; and by cutting costs, the owners have secured a large part of the world market. However, the lowering of marine standards has resulted in some episodes, including one off the shores of the coast of Canada, in which serious oil pollution would not have occurred if stricter international standards applicable to such ships had been in force.

It is quite evident, therefore, that at some time internationally agreed standards of pollution control will be required if those industries not subject to pollution restrictions are not to be allowed to acquire a dominant part of the market. It may be premature to claim that global environmental problems necessitate such a solution, but it would clearly be advantageous if the 1972 conference, called in Stockholm by the United Nations, were to be able to take some preliminary decisions on areas of pollution control on which at this point of time there might be some degree of international agreement.

chapter seven

SUMMARY AND CONCLUSIONS

1. Man is putting into the earth's atmosphere a wide variety of pollutants from different sources. These possess very different characteristics and represent multiple problems —not a single one. Continued careful measurement of pollutants in the modern city environment is necessary if control action is to be taken in time.

2. The increase in air pollution over the last forty years has been due partly to the insatiable demand for increased energy expenditure, and partly to the expanding use of the automobile for personal transport.

3. Gross particulate pollution, when combined with sulphur dioxide, has measurable effects on the materials of which cities are constructed and on the fabrics used by its inhabitants.

4. The adverse effects of air pollution on health are subtle, but they have been clearly demonstrated to be occurring in most western cities. Although the health problems created by cigarette consumption far outweigh, at the present time, those caused by air pollution, this is no reason to discount the air-pollution effects which have been documented.

5. Although the total costs incurred as a result of air pollution are bound to be crude estimates, they are, in their

totality, sufficient to necessitate a reduction of air-pollutant levels in most industrial cities.

6. Although CO, SO_2, NO, and CO_2 do not appear to be long-term problems in the earth's environment, atmospheric particles, lead, and possibly other metals released into the atmosphere do represent potential hazards. Air pollution by SST aircraft at 60,000 feet poses a special problem and an unresolved controversy.

7. One of the most important present tasks is a reconstruction of education at all levels, to bring up the new generation with an understanding of, and respect for, the natural environment.

8. Although legislation is clearly required to deal with air pollution, it raises formidable jurisdictional problems. Efforts to involve the public in policy-determination in this area have so far been rudimentary, and attempts to deal with pollution in a number of countries have illuminated the clumsiness of the decision-making process in a democratic society.

9. Sustained pressures by an informed public will continue to be required if educational systems are to be restructured and politicians kept aware of their responsibilities to the future.

APPENDIX

NATIONAL AIR QUALITY OBJECTIVES FOR CANADA

Proposed by the Hon. Jack Davis
Minister of the Environment
21 October 1971

Maximal Acceptable Levels

The maximal acceptable level corresponds to the secondary air-quality standards in the United States. It is intended to provide adequate protection against effects on soil, water, vegetation, materials, animals, visibility, personal comfort, and well-being. When this level is exceeded, control action by a regulatory agency is indicated.

Sulphur dioxide
    Annual arithmetic mean: 60 micrograms/m^3 = 0.02 ppm
    Maximal 24-hour level: 300 micrograms/m^3 = 0.11 ppm
    Maximal 1-hour level: 900 micrograms/m^3 = 0.34 ppm

Particulates
    Annual geometric mean: 70 micrograms/m^3
    Maximal 24-hour concentration: 120 micrograms/m^3

Carbon monoxide
    Maximal 8-hour level = 13 ppm
    Maximal 1-hour level = 30 ppm

Photochemical oxidants
    Annual arithmetic mean: 30 micrograms/m^3 = 0.015 ppm
    Maximal 24-hour level: 50 micrograms/m^3 = 0.025 ppm
    Maximal 1-hour level: 160 micrograms/m^3 = 0.08 ppm

Hydrocarbons
    Maximal 3-hour level: 160 micrograms/m^3 = 0.24 ppm

Maximal Desirable Levels

The maximal desirable level defines the long-term goal for air quality, and provides a basis for an anti-degradation policy for the unpolluted parts of the country and for the continuing development of control technology.

Sulphur dioxide
   Annual arithmetic mean: 30 micrograms/m^3 = 0.01 ppm
   Maximal 24-hour level: 150 micrograms/m^3 = 0.06 ppm
   Maximal 1-hour level: 450 micrograms/m^3 = 0.17 ppm

Particulates
   Annual geometric mean: 60 micrograms/m^3

Carbon monoxide
   Maximal 8-hour level =   5 ppm
   Maximal 1-hour level = 13 ppm

Photochemical oxidants
   Annual arithmetic mean: 20 micrograms/m^3 = 0.01   ppm
   Maximal 24-hour level: 30 micrograms/m^3 = 0.015 ppm
   Maximal 1-hour level: 100 micrograms/m^3 = 0.05   ppm

REFERENCES

1. An act relating to ambient air quality and to the control of air pollution (Bill C–224).
   The House of Commons of Canada.
   First Reading, 9 February 1971.
2. Air conservation.
   American Association for the Advancement of Science, Washington, D.C., 1968, 348 pages.
3. Air over Metro 1959, Winnipeg, Manitoba, Canada.
   Province of Manitoba, Environmental Health Laboratory.
4. Air pollution.
   World Health Organization, Palais des Nations, Geneva, 1961.
5. Air pollution.
   Edited by Arthur C. Stern. 3 Volumes. Second Edition.
   Academic Press, New York and London, 1968.
6. The Air Pollution Control Act, 1967.
   Statutes of Ontario.
7. Air pollution and health. A report for the Royal College of Physicians.
   Pitman Medical & Scientific Publishing Company Ltd., London, 1970, 80 pages.
8. Air pollution aspects of emission sources: cement manufacturing.
   U.S. Environmental Protection Agency, Publication No. AP–94, 1970.
9. Air quality criteria for carbon monoxide.
   National Air Pollution Control Administration, U.S. Public Health Service. Publication No. AP–62, 1970.
10. Air quality criteria for hydrocarbons.
    National Air Pollution Control Administration, U.S. Public Health Service, Publication No. AP–64, 1970.

11. Air quality criteria for nitrogen oxides.
    Air Pollution Control Office, U.S. Environmental Protection Agency, Publication No. AP–84, 1971.

12. Air quality criteria for particulate matter.
    National Air Pollution Control Administration, U.S. Public Health Service, Publication No. AP–49, 1970.

13. Air quality criteria for photochemical oxidants.
    National Air Pollution Control Administration, U.S. Public Health Service, Publication No. AP–63, 1970.

14. Air quality criteria for sulfur oxides.
    National Air Pollution Control Administration, U.S. Public Health Service, Publication No. AP–50, 1969.

15. The automobile and air pollution: a program for progress.
    U.S. Department of Commerce, Washington, D.C., 1967.

16. Bates, D. V., Bell, G., Burnham, C., Hazucha, M., Mantha, J., Pengelly, L. D., and Silverman, F.
    Problems in studies of human exposure to air pollutants.
    Canadian Medical Association Journal 103: 833-37, 1970.

17. Battle of giants confuses auto issue.
    Canadian Chemical Processing, pages 61-68, October 1970.

18. Bazell, R. J.
    Lead poisoning: zoo animals may be the first victims.
    Science 173: 130-31, 1971.

19. Biological effects of carbon monoxide.
    Annals of the New York Academy of Sciences 174: 1-430, 1970.

20. The biosphere.
    Scientific American 223: (3), September 1970.

21. Bird, P. M.
    Seminar on air pollution problems.
    Federal Control Program, Canada, 1969.

22. Broecker, W. S.
    Man's oxygen reserves.
    Science 168: 1537-38, 1970.

23. Bulletin d'Hygiène de Montréal.
    Vol. 55, No. 1, January–June 1969.

24. Chow, T. J., and Earl, J. L.
    Lead aerosols in the atmosphere: increasing concentrations.
    Science 169: 577-80, 1970.

25. City of Montreal.
    By-Law 4007, 4 March 1970.
    (Amendment to By-Law 3300 concerning air pollution.)

26. Cleaning our environment: the chemical basis for action.
    American Chemical Society, Washington, D.C., 1969.

27. Controlling air pollution in metropolitan Toronto.
    Department of Energy and Resources Management, Toronto, 1971.

28. Corporate organization for pollution control.
    Edited by R. A. Hopkinson.
    Conference Board Inc., New York, Conference Board Report No. 507, 1970.

29. Crisis: readings in environmental issues and strategies.
    Edited by R. M. Irving and G. B. Priddle.
    Macmillan of Canada, Toronto, 1971, 351 pages.

30. Dansereau, P. (ed.)
    Challenge for survival: land, air, and water for man in Megalopolis.
    Columbia University Press, 1970, 235 pages.

31. A digest of environmental pollution legislation in Canada: air and soil.
    Canadian Council of Resource Ministers, 1970.

32. Douglas, J. W. B., and Waller, R. E.
    Air pollution and respiratory infection in children.
    British Journal of Preventive and Social Medicine 20: 1-8, 1966.

33. Dustfall in the Metro Winnipeg area. 1958-1964.
    Province of Manitoba, Environmental Health Laboratory, Technical Report No. 10, 1965.

34. East, C.
    Comparaison du rayonnement solaire en ville et à la campagne.
    Cahiers de Géographie de Québec 12: 81-89, 1968.

35. East, C.
    Pollution atmosphérique et île de chaleur.
    Rapport final du projet 604–7–593 du programme de subventions sur la santé nationale, Ecole de Santé Publique, Université de Montréal, Montréal, P.Q., April 1971.

36. Economic effects of sulfur dioxide on forest growth.
    Department of Energy and Resources Management, Toronto, 1971. Reprint from Journal of the Air Pollution Control Association 21: 81-86, 1971.

37. Eisenbud, M.
    Environmental protection in the City of New York.
    Science 170: 706-12, 1970.

38. Ellis, H. T., and Pueschel, R. F.
    Solar radiation: absence of air pollution trends at Mauna Loa.
    Science 172: 845-46, 1971.

39. Environmental education: the adult public.
    Report of a Workshop Conference, Washington, D.C., 1970.
    (Sponsored by the American Institute of Biological Sciences.)

40. Environmental lead and public health.
Air Pollution Control Office, U.S. Environmental Protection Agency, Publication No. AP–90, 1971.

41. Esposito, J. C.
Vanishing air: Ralph Nader's study group report on air pollution.
Grossman Publishers, New York, 1970.

42. The global environment: M.I.T. study looks for danger signs.
Editorial: Science 169: 660-62, 1970.

43. Greenberg, D. S.
Pollution control: Sweden sets up an ambitious new program.
Science 166: 200-201, 1969.

44. Guide for air pollution episode avoidance.
U.S. Environmental Protection Agency, Publication No. AP–76, 1971.

45. Harrison, H.
Stratospheric ozone with added water vapor: influence of high-altitude aircraft.
Science 170: 734-36, 1970.

46. Hearings before the Committee on Interior and Insular Affairs, United States Senate.
17 July 1968.

47. Hearings before the Subcommittee on Air and Water Pollution of the Committee on Public Works, United States Senate.
Second Session of Air Quality Criteria, 29, 30, 31 July 1968.

48. Helfrich, H. W., Jr. (ed.)
The environmental crisis.
Yale University Press, New Haven and London, 1970.

49. Hexter, A. C., and Goldsmith, J. R.
Carbon monoxide: association of community air pollution with mortality.
Science 172: 265-67, 1971.

50. Hockin, L. E.
The British approach to the control of industrial emissions to the atmosphere.
The Canadian Mining and Metallurgical Bulletin 64: 63-67, 1971.

51. Holland, W. W., Halil, T., Bennett, A. E., and Elliott, A.
Factors influencing the onset of chronic respiratory disease.
British Medical Journal 2: 205-08, 1969.

52. Holland, W. W., and Reid, D. D.
Urban factor in chronic bronchitis.
Lancet 1: 445-48, 1965.

53. Holland, W. W., Reid, D. D., Seltser, R., and Stone, R. W.
Respiratory disease in England and the United States.
Archives of Environmental Health 10: 338-43, 1965.

54. Inman, R. E., Ingersoll, R. B., and Levy, E. A.
Soil: a natural sink for carbon monoxide.
Science 172: 1229-31, 1971.

55. An introduction to air pollution and its control in Ontario.
Department of Energy and Resources Management, Toronto,
1971.

56. Larsen, R. I.
Determining reduced-emission goals needed to achieve air
quality goals—a hypothetical case.
Journal of the Air Pollution Control Association 17: 823-29,
1967.

57. Larsen, R. I., Zimmer, C. E., Lynn, D. A., and Blemel, K. G.
Analyzing air pollutant concentration and dosage data.
Journal of the Air Pollution Control Association 17: 85-93,
1967.

58. Lave, L. B., and Seskin, E. P.
Air pollution and human health.
Science 169: 723-33, 1970.

59. Linzon, S. N.
Economic effects of sulfur dioxide on forest growth.
Journal of the Air Pollution Control Association 21: 81-86,
1971.

60. Morgan, G. B., Ozolins, G., and Tabor, E. C.
Air pollution surveillance systems.
Science 170: 289-96, 1970.

61. Mortality and morbidity during the London fog of December
1952.
London: Her Majesty's Stationery Office, Report No. 95 on
Public Health and Medical Subjects, 1954.

62. Mosher, J. C., Macbeth, W. G., Leonard, M. J., Mullins,
T. P., and Brunelle, M. F.
The distribution of contaminants in the Los Angeles Basin
resulting from atmospheric reactions and transport.
Journal of the Air Pollution Control Association 20: 35-42,
1970.

63. Nationwide inventory of air pollutant emissions.
National Air Pollution Control Administration, U.S. Public
Health Service, Publication No. AP–73, 1968.

64. Nuessle, V. D., and Holcomb, R. W.
Will the SST pollute the stratosphere?
Science 168: 1562, 1970.

65. Ontario's air pollution index.
Department of Energy and Resources Management, Toronto,
1971.

66. Petrilli, F. L., Agnese, G., and Kanitz, S.
    Epidemiologic studies of air pollution effects in Genoa, Italy.
    Archives of Environmental Health 12: 733-40, 1966.

67. La pollution de l'air par l'anhydride sulfureux et les particules
    aéroportées à Montréal.
    Rapport du Service de Santé de Montréal, 1970.

68. Pollution and our environment.
    A National Conference of the Canadian Council of Resource
    Ministers, Montreal, 1966.

69. Pollution: some questions and answers.
    Dominion Foundries and Steel, Hamilton, Ontario, 1971.

70. Rasool, S. I., and Schneider, S. H.
    Atmospheric carbon dioxide and aerosols: effects of large increases on global climate.
    Science 173: 138-41, 1971.

71. Regulations made under the Air Pollution Control Act, 1967.
    Ontario.
    General: Ontario Regulation 133/70, Sulphur Content of
    Fuels; 374/70, Air Contaminants from Motor Vehicles;
    403/68, 285/69, & 18/70.

72. Report on continuous air quality monitoring stations in Metropolitan Toronto during 1968.
    Air Management Branch, Ontario Department of Energy and
    Resources Management, 1969.

73. Report on continuous air quality monitoring stations in Metropolitan Toronto during 1969.
    Air Management Branch, Ontario Department of Energy and
    Resources Management, 1970.

74. Report on suspended particulate matter concentrations measured in Ontario during 1969 by high-volume samplers.
    Air Management Branch, Ontario Department of Energy and
    Resources Management, 1970.

75. Ross, G. H. B.
    Social accounting: measuring the unmeasurables?
    Canadian Chartered Accountant, July 1971.

76. Smoke concentration in the Metro Winnipeg area, 1957-1967.
    Province of Manitoba, Environmental Health Laboratory,
    Technical Report No. 3, 1968.

77. Van der Lende, R.
    Epidemiology of chronic non-specific lung disease (chronic bronchitis).
    Van Gorcum & Co., New York, 1969.

78. Westberg, K., Cohen, N., and Wilson, K. W.
    Carbon monoxide: its role in photochemical smog formation.
    Science 171: 1013-15, 1971.
79. Winthrop, S. O.
    Canada's Clean Air Act.
    The Canadian Mining and Metallurgical Bulletin 64: 60-62,
    1971.

INDEX